MEDICAL
INTELLIGENCE
UNIT

THE MOLECULAR BIOLOGY OF PAGET'S DISEASE

Paul T. Sharpe

University of London
London Bridge, England

Springer-Verlag Berlin Heidelberg GmbH

R.G. LANDES COMPANY
AUSTIN

MEDICAL INTELLIGENCE UNIT
THE MOLECULAR BIOLOGY OF PAGET'S DISEASE

R.G. LANDES COMPANY
Austin, Texas, U.S.A.

International Copyright © 1996 Springer-Verlag Berlin Heidelberg
Originally published by Springer-Verlag, Heidelberg, Germany 1996
Softcover reprint of the hardcover 1st edition 1996

 Springer

International ISBN 978-3-662-22507-3

While the authors, editors and publisher believe that drug selection and dosage and the specifications and usage of equipment and devices, as set forth in this book, are in accord with current recommendations and practice at the time of publication, they make no warranty, expressed or implied, with respect to material described in this book. In view of the ongoing research, equipment development, changes in governmental regulations and the rapid accumulation of information relating to the biomedical sciences, the reader is urged to carefully review and evaluate the information provided herein.

Library of Congress Cataloging-in-Publication Data

The molecular biology of Paget's disease / [edited by] Paul T. Sharpe
    p. cm. — (Medical intelligence unit)
  Includes bibliographical references and index.
  ISBN 978-3-662-22507-3   ISBN 978-3-662-22505-9 (eBook)
  DOI 10.1007/978-3-662-22505-9
  1. Osteitis deformans--Molecular aspects.   I. Sharpe, Paul T. II. Series.
    [DNLM: 1. Osteitis Deformans--physiopathology. 2. Osteitis Deformans--etiology.
3. Molecular Biology. WE 250 M718 1996]
RC931.065M65 1996
616.7'12—dc20
DNLM/DLC
for Library of Congress
        96-32383
        CIP

PUBLISHER'S NOTE

R.G. Landes Company publishes six book series: *Medical Intelligence Unit, Molecular Biology Intelligence Unit, Neuroscience Intelligence Unit, Tissue Engineering Intelligence Unit, Biotechnology Intelligence Unit* and *Environmental Intelligence Unit.* The authors of our books are acknowledged leaders in their fields and the topics are unique. Almost without exception, no other similar books exist on these topics.

Our goal is to publish books in important and rapidly changing areas of the biosciences for sophisticated researchers and clinicians. To achieve this goal, we have accelerated our publishing program to conform to the fast pace in which information grows in the biosciences. Most of our books are published within 90 to 120 days of receipt of the manuscript. We would like to thank our readers for their continuing interest and welcome any comments or suggestions they may have for future books.

Shyamali Ghosh
Publications Director
R.G. Landes Company

DEDICATION

This book is dedicated to the memory of Jean Stevens, founder of the Salford Paget's Appeal, who died in June 1995.

CONTENTS

=EDITORS=

Paul T. Sharpe
Department of Craniofacial Development
United Medical and Dental Schools of Guy's and St. Thomas's Hospitals
University of London
London Bridge, England
Chapter 6

= CONTRIBUTORS =

David C. Anderson
Hong Kong
Chapter 2

R. John Barr
Department of Orthopaedic Surgery
Musgrave Park Hospital
Belfast, Northern Ireland
Chapter 8

Mark A. Birch
Human Bone Cell Research Group
Department of Human Anatomy
  and Cell Biology
The University of Liverpool
Faculty of Medicine
Liverpool, England
Chapter 5

Anne Demulder
University of Texas Health Science
  Center at San Antonio
San Antonio, Texas, U.S.A.
Chapter 3

Anthony J. Freemont
Department of Rheumatology
The University of Manchester
Manchester, England
Chapter 1

James A. Gallagher
Human Bone Cell Research Group
Department of Human Anatomy
  and Cell Biology
The University of Liverpool
Faculty of Medicine
Liverpool, England
Chapter 5

Anne E. Hughes
Department of Medical Genetics
The Queens University of Belfast
Belfast, Northern Ireland
Chapter 8

Linda M. McManus
University of Texas Health Science
  Center at San Antonio
San Antonio, Texas, U.S.A.
Chapter 3

Andrew P. Mee
Department of Medicine
Manchester Royal Infirmary
Manchester, England
Chapter 4

Michael J. Rogers, Ph.D.
The University of Sheffield Medical
  School
Department of Human Metabolism
  and Clinical Biochemistry
Sheffield, England
Chapter 7

G. David Roodman
University of Texas Health Science
  Center at San Antonio
San Antonio, Texas, U.S.A.
Chapter 3

R. Graham G. Russell, M.D., Ph.D.
The University of Sheffield Medical
  School
Department of Human Metabolism
  and Clinical Biochemistry
Sheffield, England
Chapter 7

PREFACE

Paget's disease of bone is a "Cinderella" of human diseases. Although it can be severely debilitating and affects around 5% of the elderly population of Europe and the United States it receives scant recognition from clinicians, pharmaceutical companies and the general public. Add to this fact that the disease is probably caused by a viral infection and can occasionally lead to such fatal complications as osteosarcoma, it is remarkable that Paget's disease appears to attract so little attention. There is a great deal of ignorance in the general medical profession regarding the diagnosis and treatment of Paget's disease and the two major Paget's disease charities—the Paget's Foundation, USA and the National Association for the Relief of Paget's Disease, UK—which are very active in educating physicians and patients. An excellent monograph on the pathophysiology of Paget's Disease by Professor John Kanis, M.D., (University of Sheffield, UK) was published in 1991. The aim of this new text is to complement that text by concentrating on areas not covered by John Kanis, namely the molecular and cellular rather than clinical aspects of the disease.

Leading researchers in different aspects of Paget's disease biology have contributed a total of eight chapters that together summarize the current status of knowledge and thinking. The book is written for both researchers and physicians and aims to stimulate further research into the fascinating disease.

The book begins with a chapter of the basic pathology of the disease.

Chapter 2 details the highly unusual epidemiology and introduces the idea of a viral cause.

Chapter 3 describes the cell and molecular biology of Pagetic osteoclasts, the cells at the heart of the disease which are responsible for the initial lesion.

In chapter 4, the viral etiology is discussed in great detail. Chapter 5 discusses the role of cytokines and growth factors and is followed in chapter 6 by the piecing together of many of the molecular findings into the first molecular model of the biology of Pagetic bone cells.

Chapter 7 concerns the mechanism of the action of the drugs used to treat the disease, and the book ends with an insight into a fascinating inherited form of Paget's Disease, Familial Expansile Osteolysis.

INTRODUCTION

PAGET'S IS AN *OLD* DISEASE

The condition now known as Paget's disease was first reported by Wilks in 1869[1] but it was a few years later before the disease was classically described as *Osteitis deformans* (Paget's disease) by Sir James Paget in 1877 working in London.[2] The disease itself is certainly much older, with pagetic lesions being found in skeletons from ancient Egypt,[3] the Gallo-Roman era[4] and Anglo Saxon burial grounds of around AD950.[5] Paget's is primarily a disease of the elderly affecting between 5-8% of the population over 40 in Europe and the United States.[6]

PAGET'S IS AN ENIGMATIC DISEASE

While the scientific literature is increasingly filled with great excitement of genes being identified causing diseases that affect a few thousand people, Paget's, a debilitating disease that may affect up to five million people worldwide, receives scant attention. There is no single reason for this anomaly but a combination of several features that together make Paget's a "Cinderella" of diseases. *Paget's disease is not Mendeally inherited and so there is not a single "Paget's disease gene" to be searched for. Paget's is primarily a disease of the elderly population. Paget's disease is non-infectious. Excellent new treatments are available that if not cure, certainly alleviate much of the suffering. Paget's disease is rarely lethal.* Balanced against these features are the inescapable facts that *Paget's is often very a painful, severely debilitating disease that can lead to serious incurable complications such as osteosarcoma. Paget's disease is a "Western" disease with a unique epidemiology that has no proven cause despite being first described over 100 years ago.*

PAGET'S IS NOT AN IDIOSYNCRATIC DISEASE

Paget's disease research has a great deal to offer the bone field: a fact that has not been fully appreciated by researchers, pharmaceutical companies and funding bodies alike. Unlike "normal" bone and bone in diseases such as osteoporosis, pagetic bone is very cellular. It is a very active tissue with many osteoclasts and osteoblasts carrying out their functions in an uncontrolled fashion. So much basic bone research has been carried out with cultured cells that the bone field often feels like an unreal world where the tissue itself is forgotten. The search for the mysterious "coupling factor" that communicates osteoblasts and osteoclasts still occupies much of the basic research on bone. In pagetic tissue, coupling is abnormal and can provide a unique environment to study basic bone biology in vivo.

PAGET'S IS NOT A CONTROVERSIAL DISEASE

The only major controversy is whether or not it is caused by a virus infection. Sadly the controversy ends there because if viral infection is not the cause there are no feasible alternative ideas. This book is not meant to be an all encompassing work on Paget's disease but rather to highlight the molecular and cell biology of the disease. The aim of this book is to provide a forum for top researchers in the field to be controversial and express their own ideas. From the onset I have encouraged all authors to be speculative. In truth there is so little hard information on the molecular biology of Paget's disease that almost any speculation can fit with what is known. There is some degree of duplication between chapters which I have deliberately chosen not to edit so that authors' different interpretations can be appreciated. Above all I hope this book will stimulate new ideas.

REFERENCES
1. Wilks S. Case of osteoporosis or spongy hypertrophy of the bones (calvaria, clavice, os femoris and rib exhibited at the society). Transactions of the Pathological Society of London 1869; 20:273-277.
2. Paget J. On a form of chronic inflammation of bones (osteitis deformans). Medico-Chirurgical Transactions of London 1877; 60:37-63.
3. Hutchinson J. On osteitis deformans. Illustrated Medical News 1889; 2:169- 180.
4. Astre G. Maladie osseuse pagétoide d'un gallo-romain. Revue Pathologie Genérale et Comparée 1957; 57:955-961.
5. Wells C, Woodhouse NJY. Paget's disease in an Anglo-Saxon. Medical History 1975; 19:396-400.
6. Kanis JA. Pathophysiology and Treatment of Paget's Disease of Bone. London: Martin Dunitz Ltd, 1991.

FURTHER INFORMATION ON PAGET'S DISEASE CAN BE OBTAINED FROM:

National Association for the Relief of Paget's Disease
1 Church Road
Eccles
Manchester M30 0DL, UK
Tel. 0161 707 9225.

The Paget Foundation
200 Varick Street - Suite 1004
New York, NY, USA 10014-4810
Tel. 212 229 1592
Fax. 212 229 1502.
Email. pagetfdn@aol.com

ACKNOWLEDGMENTS

My thanks go to all the contributors for making the editing of this book an enjoyable and interesting experience. My personal special thanks also go to the three charitable organizations who have provided and hopefully will continue to provide invaluable support for all kinds of matters relating to Paget's disease. The National Association for the Relief of Paget's Disease, The Paget Foundation (USA) and the Salford Paget's Appeal. Thank you.

CHAPTER 1

THE PATHOLOGY OF PAGET'S DISEASE

Anthony J. Freemont

INTRODUCTION

Paget's disease[1] is characterized by excessive, uncoordinated, bone turnover resulting in disorganization of bone architecture, manifested most commonly as thickening, increased bone mass and deformity. Although commonly discussed in texts with the metabolic or generalized bone diseases the disease typically affects part or all of one, several or many bones, but never all the bones and usually only one or two. This simple, and frequently ignored fact, is the single most perplexing and frustrating aspect of the disease when consideration is given to the mechanisms underlying its pathogenesis.

CLINICAL FEATURES

The onset of Paget's Disease is insidious. Typically most patients (>90%) with the disorder are asymptomatic.[2] The major presenting complaints are bone pain and symptoms related to fracture and deformity of bone.[3] These symptoms result, usually indirectly, from the characteristic changes that occur in bone cell metabolism. In this chapter the pathological appearances of affected bone will be used to link clinical symptomatology to disturbed cell function.

The Molecular Biology of Paget's Disease, edited by Paul T. Sharpe.
© 1996 R.G. Landes Company.

The extent and nature of the symptoms and overall morbidity of the disease are directly linked to: (a) the number of bones involved; (b) the degree of bone enlargement and deformity; (c) which bones are involved; and (d) the length of time from onset of the disorder.[4]

GENERAL PATHOPHYSIOLOGY

To understand Paget's disease requires an understanding of normal bone.

NORMAL BONE

A bone is a complex organ.[5] It has two major components—the bony material itself and bone marrow. Both are altered in Paget's disease.

Bone, as opposed to *a* bone, has two constituents, the matrix (made up of layers of collagen fibers and associated molecules) and four types of bone cell (two transient cells; osteoclasts, which resorb bone, and osteoblasts, which build bone; and two fixed cells, both of which are osteoblast progeny, resting surface cells and osteocytes).

1. The matrix

Matrix is the main referent of bone. Within a specific bone the bony matrix usually has a characteristic distribution. The outer part of the bone consists of a shell of bone matrix, called the cortex, or cortical bone. Generally it forms a dense box or tube that gives the overall conformation or shape to the bone. Within this box lie thin struts of matrix material called trabeculae which have a general orientation along the lines of force acting through that bone (Fig. 1.1).

The cortical bone is divided into three zones from the outside inwards (Fig. 1.2).

The outer zone is called periosteal bone. Depending upon which bone is examined, periosteal bone accounts for between one-tenth and one-fifth of the thickness of the cortex. It is made by a layer of cells that cover the bone surface, called periosteal cells. These

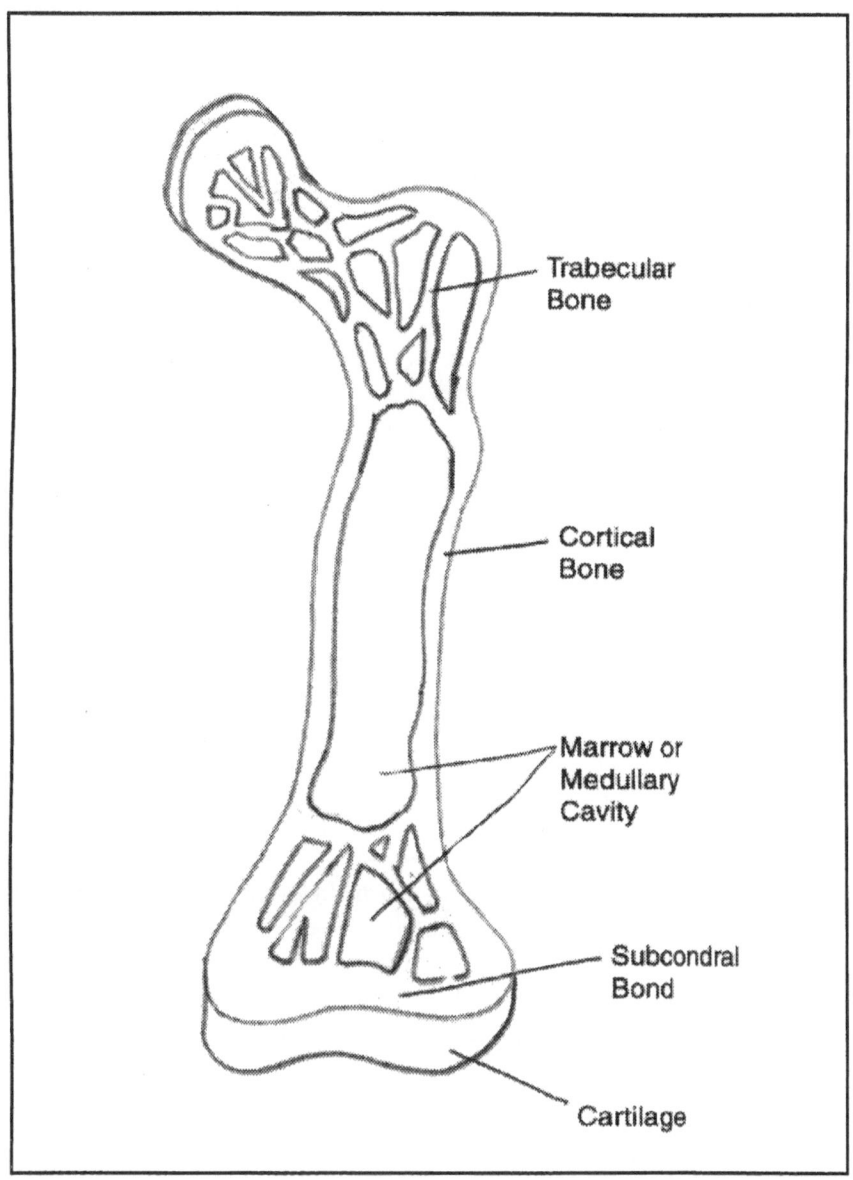

Fig. 1.1. A diagram showing the relationship between cortical and trabecular bone and bone, cartilage and marrow.

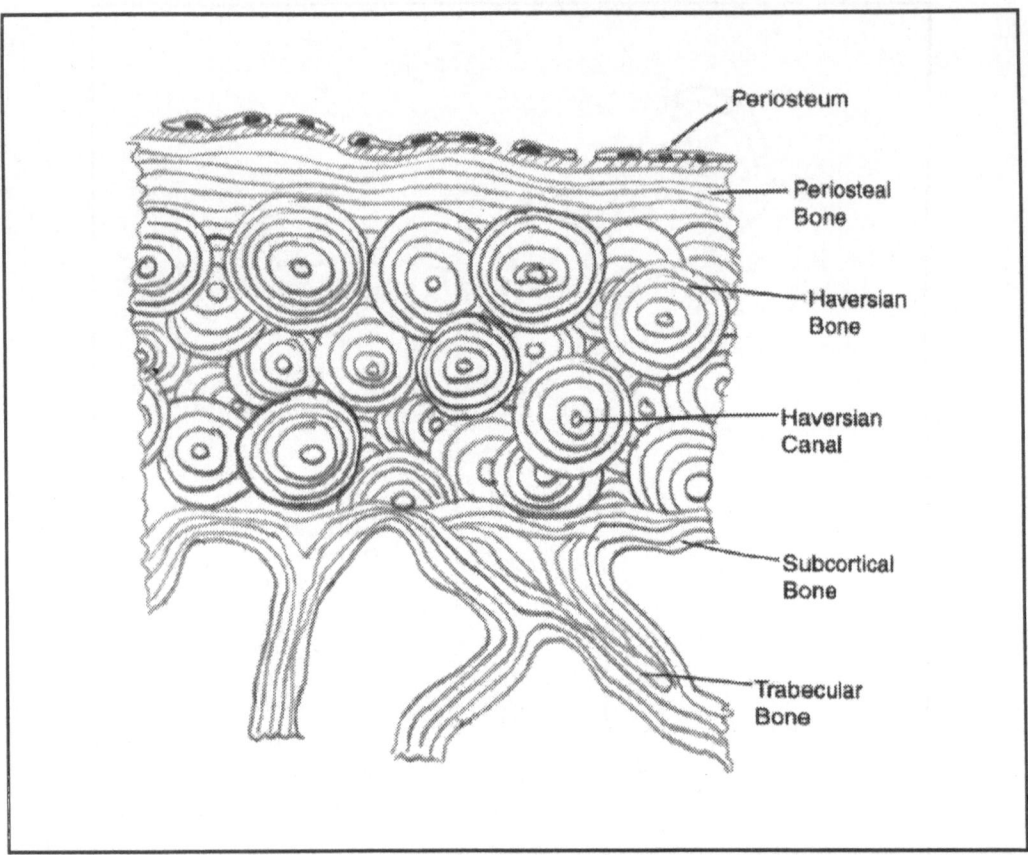

Periosteum

Periosteal
Bone

Haversian
Bone

Haversian
Canal

Subcortical
Bone

Trabecular
Bone

Fig. 1.2. A diagram of the various layers of cortical bone and their relationship to other components of bone.

cells are distinct from the osteoblasts within bone and are not subject to the same environmental controls. Periosteal bone is deposited as a shell around the bone continuously throughout life at a more or less steady rate. Under normal circumstances, however, the layer of periosteal bone does not get thicker, because it is being continuously changed into the main type of cortical bone, Haversian bone, from its inner aspect.

The majority of the cortex is made up of "Haversian" bone, constructed from interlocking, eroded spindle-shaped bone units, each with a vessel running through the center of the long axis in a fine canal called the Haversian canal. New Haversian systems are formed by resorption of a characteristically shaped space within existing Haversian or periosteal bone, followed by ingrowth of vessels and deposition of new bone. The process of Haversian canal formation is seen wherever thick or dense bone forms and may well be a mechanism to bring nutrients to osteocytes deep within the matrix. In pathological conditions, such as Paget's disease, where bone matrix comes to occupy a significant proportion of the total tissue volume, Haversian systems based around Haversian canals, develop in trabecular as well as cortical bone.

On the inner, or marrow surface, of the Haversian bone there is a thin layer of subcortical bone. The spicules of trabecular bone arise from the subcortical bone and form a meshwork of strands and plates of bone within the marrow cavity (Fig. 1.3).

Bone matrix has four major molecular components, Type 1 collagen, noncollagenous proteins, nonproteinaceous organic material, and a crystalline compound of calcium, phosphorous and oxygen which, in its crystal structure most closely resembles the compound hydroxyapatite.

Collagen is the most abundant organic molecule in bone matrix. It consists of long fibrils which in bone are aggregated to form thin rafts, a few micrometers thick, known as lamellae. In

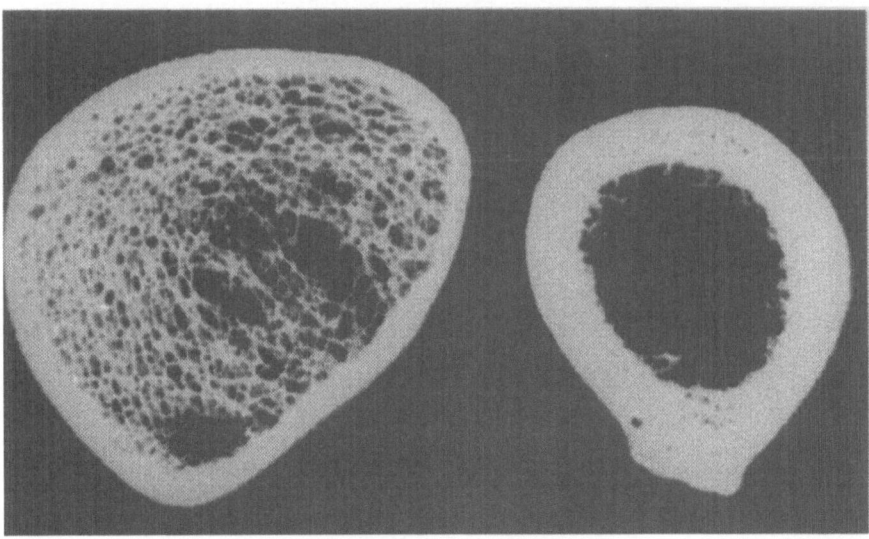

Fig. 1.3. Slab radiograph of a transverse section of the mid shaft and lower shaft of the femur showing the relationship between trabecular and cortical bone. x0.5.

each lamella the long axes of the collagen fibers are parallel. The long axes of the collagen fibers are at right angles in adjacent lamellae. At its most basic level, therefore, bone matrix is a laminate. Bone formed in this way is known as lamellar bone and is very strong (Fig. 1.4).

The noncollagenous organic molecules are closely and regularly associated with the collagen fibers and their distribution and quantity are regulated by the close packing and orientation of the collagen fibers. As several are responsible for binding the crystalline hydroxyapatite to the organic matrix, their quantity (and thus indirectly the collagen fiber number and disposition) are responsible for the amount of hydroxyapatite per unit volume of bone. In certain pathological states, including Paget's disease, the normal lamellar structure of bone may be lost and the collagen fibers have a near random distribution, resembling the weave of a crude cloth. This bone is known as "woven" bone. The less dense pack-

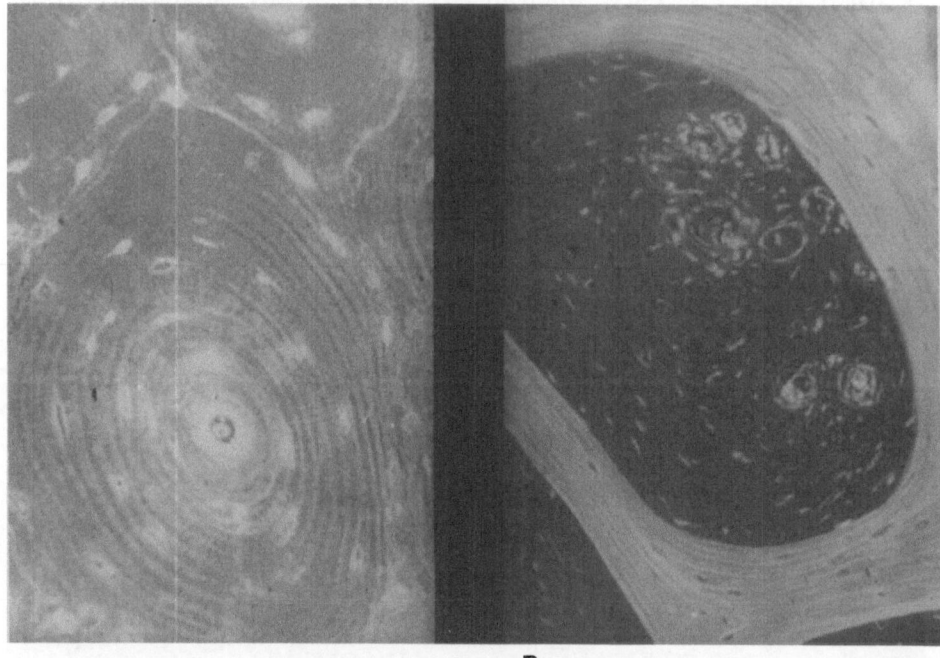

A   B

Fig. 1.4. Polarizing micrograph of Haversian cortical (A) and trabecular bone (B). Both are lamellar but the orientation of the collagen fibers is different in the different types of bone. x150.

ing of the randomly organized collagen fibers means that the volume of the bone per unit amount of collagen is increased. It follows from the disorganized packing of collagen fibers, that the noncollagenous matrix molecules and the amount of calcium are increased. Woven bone is therefore bulkier and more densely calcified and, consequent upon the loss of collagen fiber orientation, weaker.

2. Bone cells

Intimately associated with the matrix are four "bone cells"— osteoblasts, which form bone on bone surfaces; osteoclasts, which resorb bone, again usually on bone surfaces; osteocytes, the cells embedded within bone matrix, that may have a mechanoreceptor function; and resting surface cells that cover all bone surfaces other than the 15% or so covered by osteoblasts and osteoclasts (Fig. 1.5).

The nature and amount of the bone matrix is determined by the balanced activity of the four cell types. In particular, it is usual for the amount of bone being deposited by osteoblasts to be equal to that being resorbed by osteoclasts, thus maintaining the total amount of bone constant. Although it is not entirely clear exactly how the activity of these two types of bone cell is coordinated the linkage, or coupling, between the two is known to be essential for maintenance of normal bone structure.[6]

3. Bone marrow

Within a bone, the space not occupied by bone and its cells is filled by marrow. The major components of marrow are fat and hemopoietic cells, but there are also fibroblasts, osteoprogenitor cells and blood vessels.

BONE CHANGES IN PAGET'S DISEASE

Every one of the elements of bone, matrix, bone cells, and bone marrow, is altered in Paget's disease.

In Paget's disease the alteration in bone cell and bone matrix biology is very complex and varies between patients, between bones and within a bone. In general terms the three main changes are: (a) an increase in osteoclasis; (b) disturbed bone cell coupling; and (c) alterations in matrix synthesis and turnover.

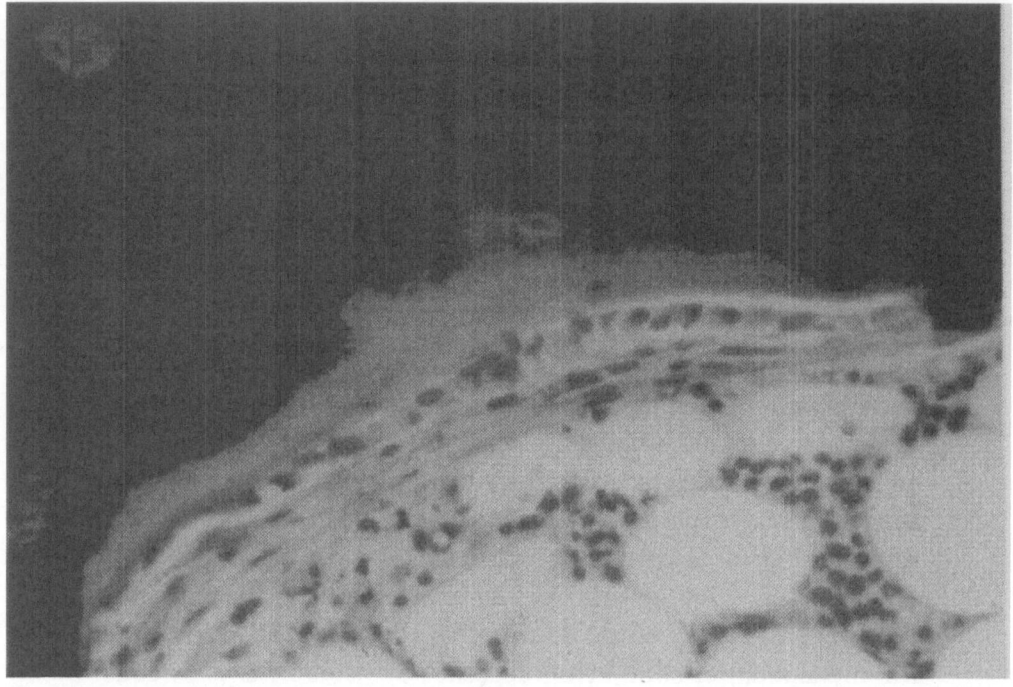

A

B

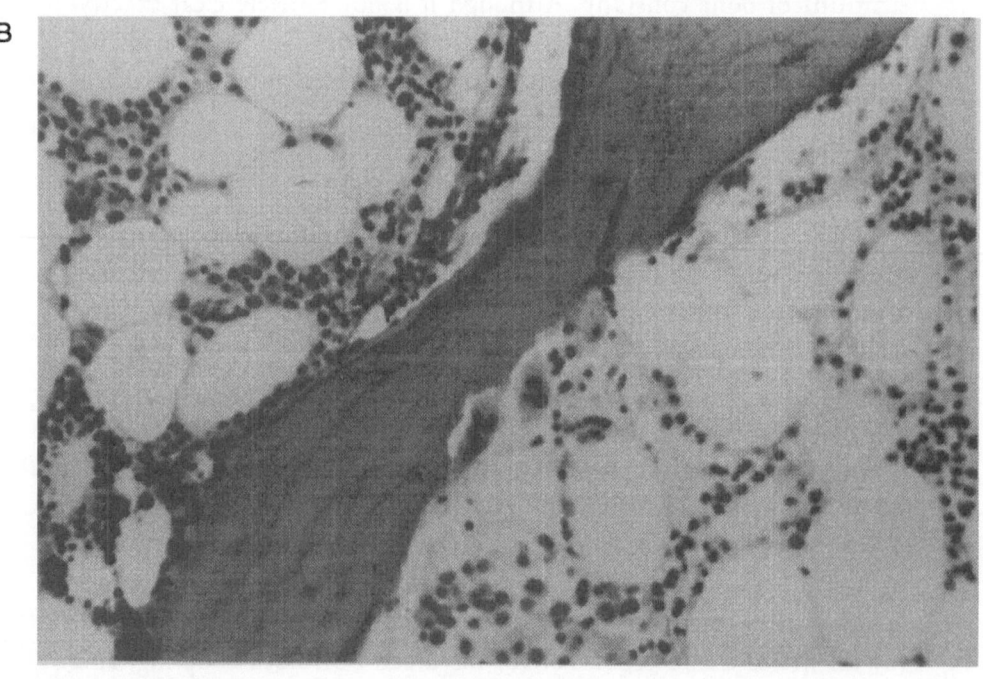

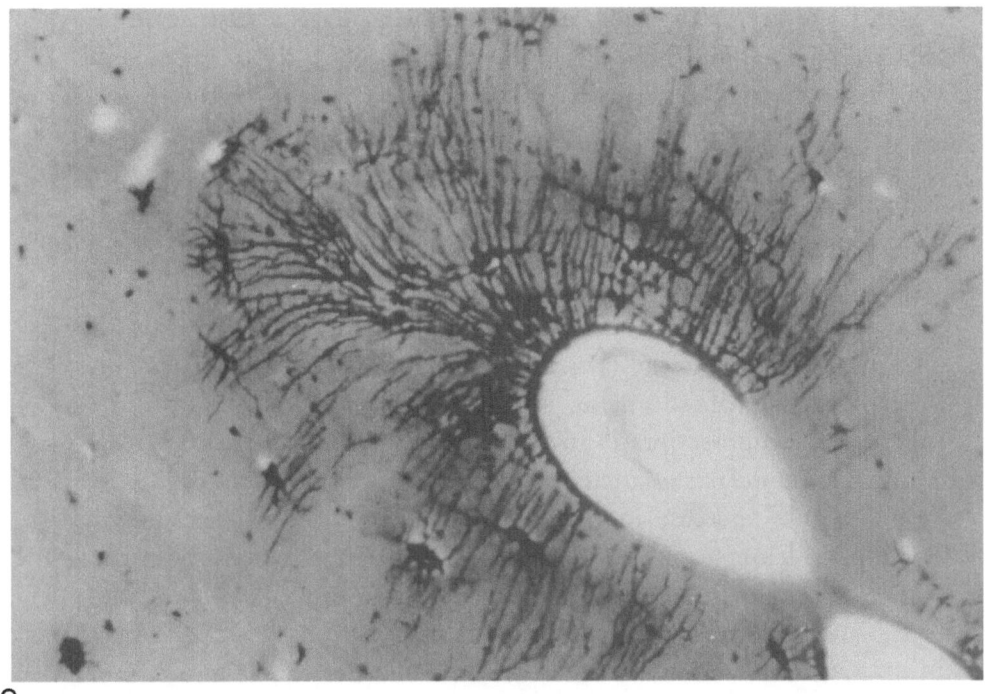

C

D

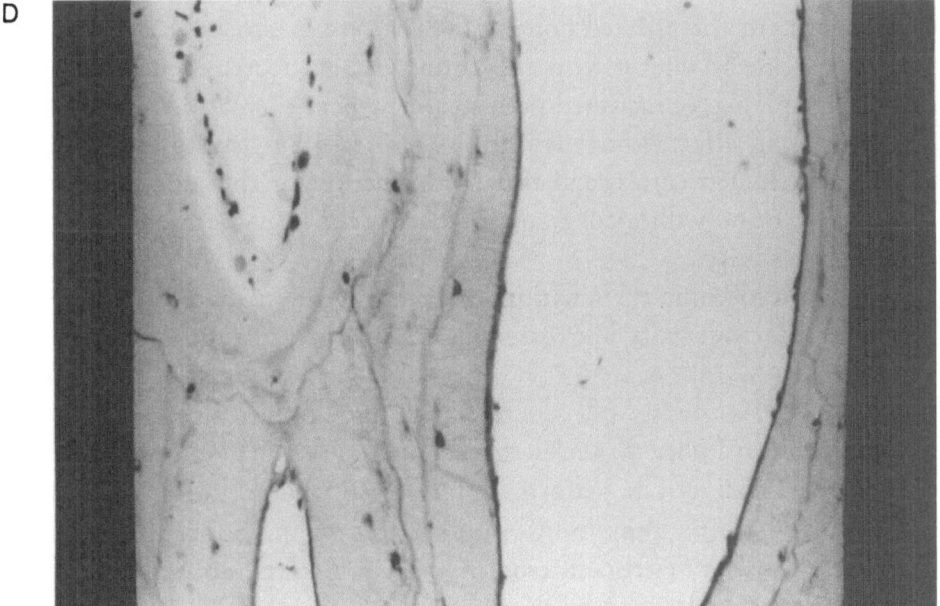

Fig. 1.5. A montage of the 4 types of bone cells. (A) Osteoblast x200, (B) Osteoclast x150, (C) Osteocyte x100 and (D) Resting surface cell x100.

There are two major manifestations of Paget's disease recognized on imaging. Lytic Paget's disease in which the total amount of bone matrix is decreased and sclerotic in which it is increased.[7]

The lytic variant of Paget's disease, is much rarer than the sclerotic form at the time of diagnosis.[8] In the lytic variant there is a net loss of bone, which may be an isolated phenomenon or be associated with the leading edge of an area of sclerosis. It is, in general, seen most commonly in two situations—in younger people, with presumed earlier disease; and in certain bones, notably the skull. The skull is the classical site for the lytic area to be at the leading edge of advancing sclerosis, a pattern known as "osteoporosis circumscripta".[9] In lytic Paget's disease the number and size of osteoclasts is increased but there is either no or a lesser comparable increase in osteoblastic activity. This disturbance in coupling leads to a loss of bone. The pattern of activity is such that the excessive osteolysis starts as a focus in the bone and then spreads away from that focus.

Osteosclerosis, or an increase in bone density, is much more typical of Paget's disease at the time of its presentation. In this form of the disease there is an increase in the total amount of matrix within the affected bone. When viewed macroscopically an affected bone is thicker than the equivalent unaffected bone; this is a consequence of increased periosteal new bone formation. When transected an affected bone is not only enlarged but is also seen to have a thickened cortex, and usually an increase in the amount of trabecular bone with a concomitant decrease in the volume of the medullary cavity.

On microscopic examination the picture is dominated by the number of osteoclasts and osteoblasts and the amount of bone matrix.[10]

Osteoclasts are not only increased in number but also in their volume, the number of nuclei they contain and in the amount of resorption each cell is performing (Fig. 1.6). In addition osteoclasts may be seen away from bone surfaces within bone marrow, within medullary vessels and eroding osteoid, features not displayed by normal osteoclasts. Osteoclastic activity, particularly in the least severely affected areas, is greater in cortical bone than it is in trabecular bone.

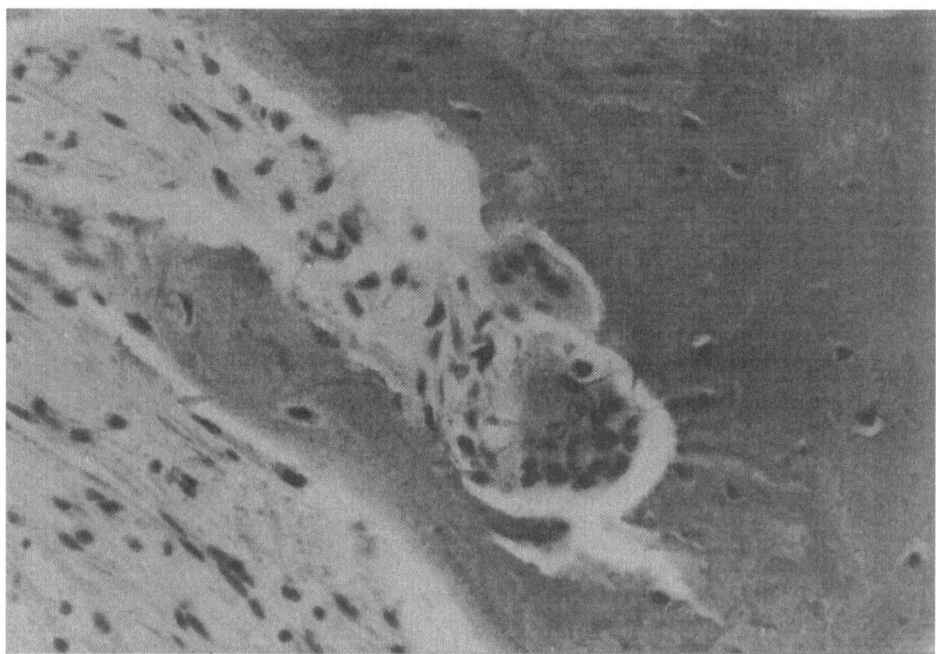

Fig. 1.6. An osteoclast from pagetic bone. It is much larger than normal and shows very active erosion. x400.

One aspect of the ultrastructure of osteoclasts has stimulated considerable research and greater controversy in recent years.[11,12] When viewed with the electron microscope osteoclasts from patients with Paget's disease are seen to contain cytoplasmic and nuclear inclusions not seen in normal osteoclasts (Fig. 1.7). At high magnifications the inclusions are seen to consist of regular arrays of particles with the overall morphology and arrangement of fully formed viral particles with features suggestive of the paramyxoviruses.[13] These are discussed in greater detail in subsequent chapters.

Increased osteoclastic activity is marked biochemically by a raised urinary hydroxyproline and collagen crosslinks.

Osteoblastic activity is generally increased and osteoblasts are of increased size. In situ studies (both immunohistochemistry and in situ hybridization) have shown the increased size to be associated with increased biosynthesis of both matrix molecules, such as type I collagen, and regulatory cytokines, such as IL-6.[14] In addition

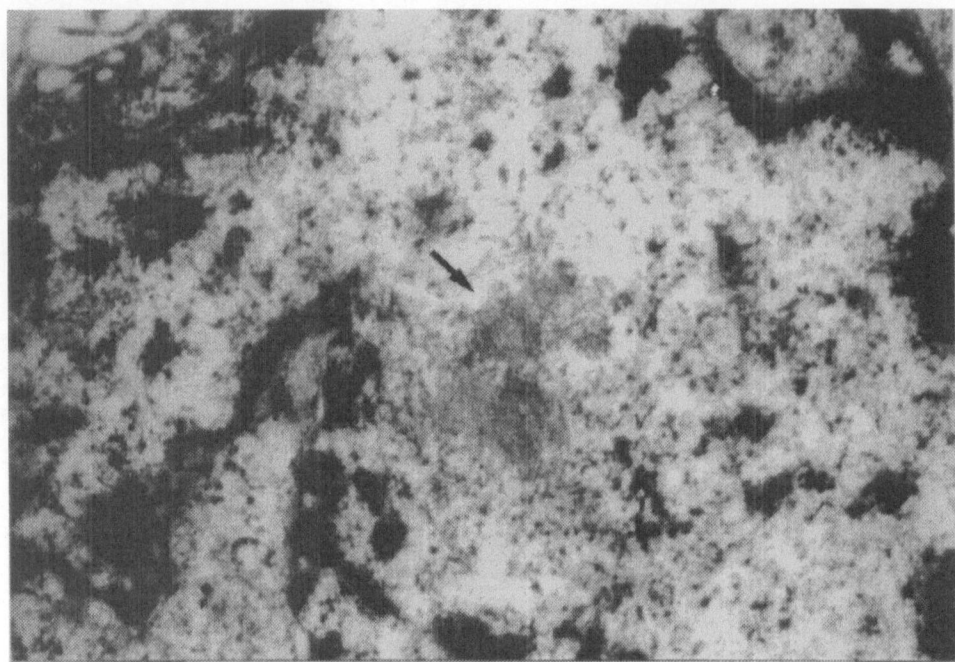

Fig. 1.7. An osteoclast nucleus from pagetic bone. It contains unusual inclusions now believed to be viral particles (arrowed).

the absolute number of osteoblasts is raised and consequentially the total amount of bone synthesis is increased. The increased osteoblastic activity is marked by an increase in the serum levels of bone specific alkaline phosphatase.

In Paget's disease the total volume of nonmineralized osteoid in both trabecular and cortical bone is raised substantially as is the proportion of mineralized bone surfaces covered by osteoid seams (Fig. 1.8). The thickness of each of the osteoid seams is usually within the accepted normal range, although towards its upper limit, and most are covered by osteoblasts. It is not surprising therefore to find that the rate at which the osteoid is mineralizing (the appositional rate) is either normal, or increased, as assessed by double tetracycline labeling (Fig. 1.9). The hyperosteoidosis is not, therefore, a manifestation of a mineralization defect (osteomalacia) but markedly increased matrix synthesis. As much of the matrix has a lamellar pattern it can be stated that in

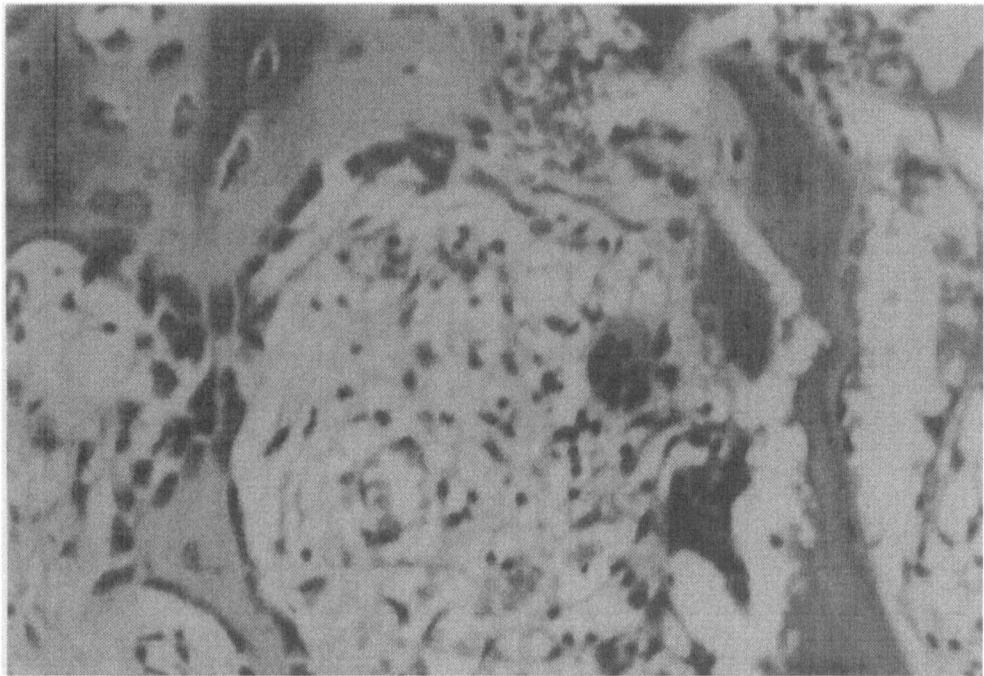

Fig. 1.8. A view of pagetic bone. There are increased numbers of osteoblasts and osteoclasts, hypervascular bone marrow, woven bone (some of which is nonappositional), and an increase in the amount of pale staining osteoid. x9,000.

most cases of Paget's disease osteoblastic activity is qualitatively normal, but quantitatively increased.

This is not always the case as the osteoblasts sometimes synthesize woven bone, particularly in the most active disease, which in this context should be regarded as evidence of a qualitative dysfunction of the osteoblast. The question of whether the matrix in Paget's disease is lamellar or not is, however, surprisingly controversial. In all but the most severe forms of the disease the matrix is clearly lamellar, but as the most severe forms tend to be the most studied and they are dominated by woven bone, the impression could be obtained that this is the predominant pattern of collagen fiber distribution in bone in Paget's disease. The situation is made difficult to assess because of the general increase in bone remodeling. Osteoclastic bone resorption and subsequent bone deposition are so active that the bone is rapidly broken up into small interlocking blocks of bone resembling a mosaic (Fig. 1.10).

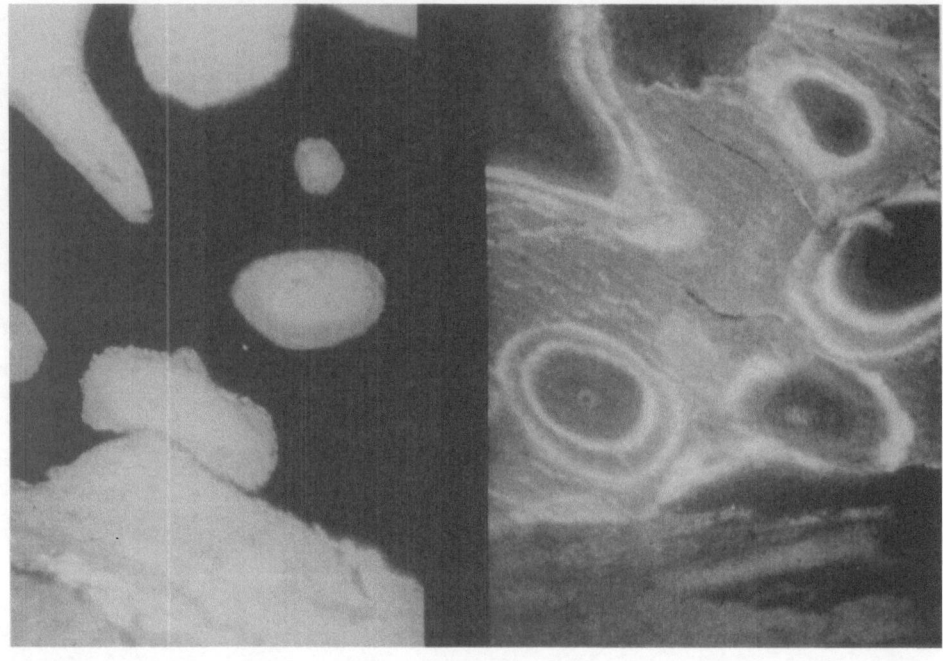

A   B

Fig. 1.9. Cortical bone stained by von Kossa's technique to demonstrate mineral (A) and a serial unstained section viewed in ultraviolet light to show tetracycline labeling (B). The distance between the labels is 50% greater than usual demonstrating the increased appositional rate in Paget's disease. x100.

The blocks can be so small and haphazardly arranged that the small fragments of interlocking lamellar bone can be mistaken by the unwary as having a woven pattern. This arrangement has been described as "pseudo-woven" (Fig. 1.11). Nevertheless in the most severe forms of the disease appositional bone formation by osteoblasts (i.e., that occurring on previous existing bone) may be woven rather than lamellar. This disruption of collagen fiber orientation makes the bone bulkier per unit volume of collagen, and weaker. In addition, again in the more severe forms of the disease, nonappositional bone may form (Fig. 1.8). This is bone forming de novo within marrow. This bone is not laid down by groups of osteoblasts but is deposited around individual osteoblast like cells. It too is woven.

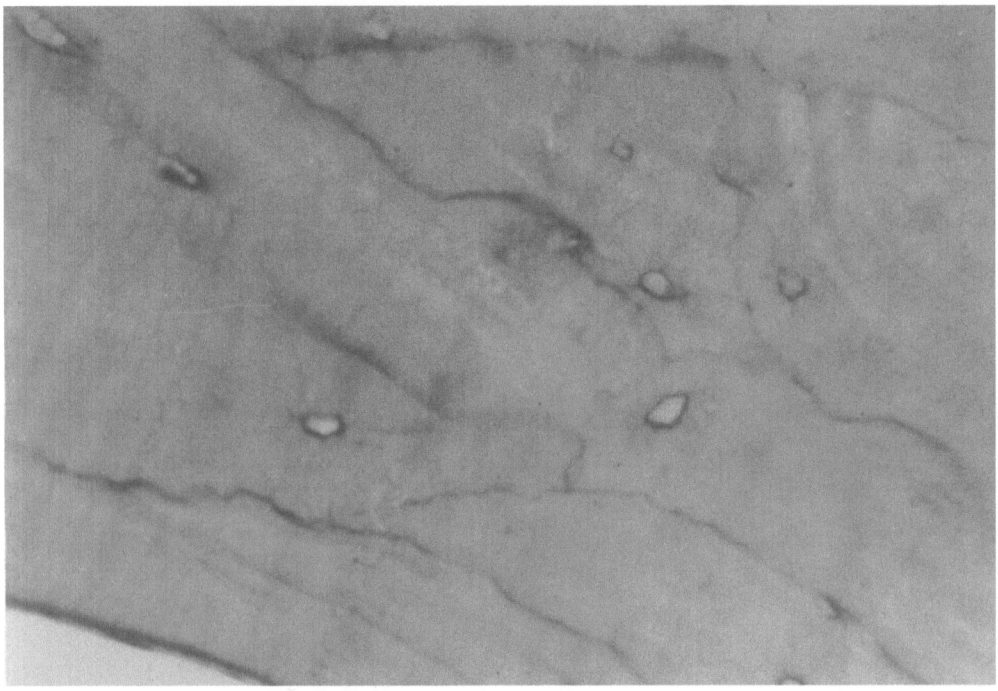

Fig. 1.10. In this section stained to show cement lines the "mosaic pattern" of bone segments is clearly seen. x75.

As the amount of bone produced increases there is a dramatic increase in the amount resembling Haversian cortical bone. This comes from an exaggeration of the normal process of "haversionalization" of periosteal cortical bone and the "corticalization" of the increased amount of subcortical and trabecular bone. This "corticalized" bone is very dense, which explains some of the radiological features seen in Paget's disease. The increased bone density seen on X-ray can also be explained by the change from lamellar to woven bone. Although woven bone contains less collagen per unit volume, the concentration of calcium is greater. Thus the amount of the radiodense calcium salts per unit volume of bone is increased, in addition to the absolute volume being increased (Fig. 1.12).

The bone marrow, which normally consists of a mixture of fat cells and hemopoietic tissue, becomes replaced with fibrous tissue.

Fig. 1.11. A section of pagetic cortical bone viewed in polarized light. To the left the collagen fibers are deposited in a typical lamelar pattern, whilst on the right they have a "pseudowoven" pattern. x50.

Initially this process is localized to bone surfaces but eventually fills the marrow cavities (Fig. 1.8). The marrow is also hypervascular, with an increase in the number and size of blood vessels when compared with normal marrow.

Unaffected bones in patients with Paget's disease may be normal or, more commonly, show a general increase in normal bone cell activity with an increase in osteoclastic and osteoblastic activity. The osteoclasts are of normal size, but of increased number. Osteoblastic activity appears physiological in nature but increased in amount.[15] The inference is that, in Paget's disease, there are humoral factors with the potential to stimulate bone cells.

COMPLICATIONS

The most common symptom in patients with Paget's disease is bone pain. In addition there is an increased incidence of fracture through the affected bone, nerve compression, bone cancer,

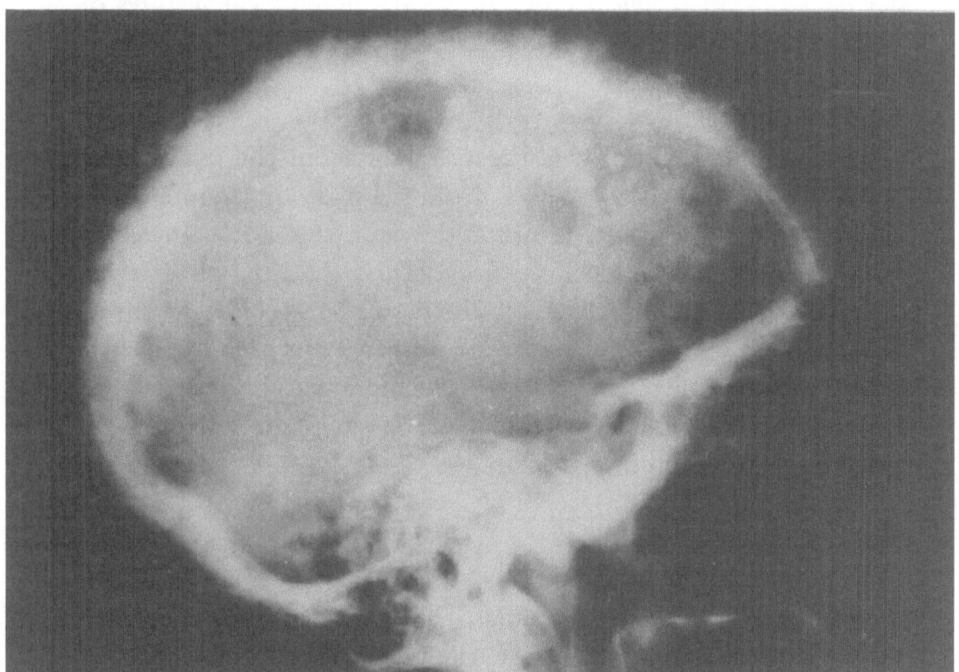

Fig. 1.12. A skull X-ray showing the increased radiodensity of pagetic bone. x0.25.

osteoarthritis, spinal ankylosing hyperostosis and, it is said, heart failure.[16,17]

PAIN

The source of bone pain is not known, but may well arise from disruption of the highly innervated periosteum. In addition the marrow, and particularly the walls of its vessels, contain nerve fibers. It has been suggested that some part of the bone pain in Paget's disease and other bone disorders in which there is increased blood flow comes from stimulation of these nerves.

FRACTURE

The increased risk of fracture of pagetic bone is thought to be due, paradoxically, to the decreased strength under load of the thickened bone which is consequent upon its disturbed collagen fiber orientation. Interestingly, fracture healing tends to be normal but the late remodeling fracture shows the histological features of Paget's disease.

NERVE COMPRESSION

Nerve compression occurs most commonly in the skull where nerves penetrate the skull to pass from the brain or brain stem to gain access to the tissues covering the skull and neck. Increased periosteal new bone formation also occurs in the edges of these ostia where the penetrating nerve fibers exit bones. The cross sectional area of the ostia is decreased and nerve compression results. One common consequence of nerve compression is deafness due to compression of the eight cranial nerve within the narrowed internal acoustic meatus.

NEOPLASIA

The primary bone neoplasms associated with Paget's disease are primarily osteogenic sarcomas. They may or may not form osteoid and in those that do not, the phenotype of the tumor cell can be determined by its expression of alkaline phosphatase, a marker of bone forming cells.[18]

There is said to be a greater incidence of secondary neoplasms within pagetic bone than within normal bone. This may be a reflection of the greater blood flow through pagetic bone than normal bone or may result from increased expression of oncogenes such as cFOS (chapter 6).

OSTEOARTHRITIS

Osteoarthritis is an end stage disorder of cartilage in which there is altered cartilage matrix turnover secondary to disturbed chondrocyte function. In Paget's disease this is thought to be secondary to the altered mechanics of the affected subchondral bone.

ANKYLOSING HYPEROSTOSIS

Ankylosing hyperostosis is a form of spinal disease where large outgrowths of bone (osteophytes) form at vertebral margins and cross the disc space between the vertebrae, fusing the spine. When associated with Paget's disease pagetic changes are present in the vertebral bodies and in the associated osteophytes.

HEART FAILURE

The increased risk of heart failure is said to be caused by increased strain on an elderly heart caused by the increased blood flow through the hypervascular bone.

SUMMARY

Paget's disease is a fundamental, but localized, dysfunction of bone cells resulting in disturbed structure and function of individual bones. All bone cells are affected in the disorder and, in addition, there is a field change within the marrow resulting in stimulation of stromal precursors and inhibition or dispersal of hemopoietic cells.

Careful consideration of the pathology of the bone poses several questions about the pathophysiology of this disorder that can only be answered by a better understanding of the nature of the disturbed cell and molecular biology in the disease. These studies are the basis of the remaining chapters in this book.

REFERENCES

1. Paget J. On a form of chronic inflammation of bones (osteitis deformans). Trans Med Chir Soc Lond 1877; 60:37-63.
2. Dalinka MK, Aronchick JM, Haddad JG. Paget's disease. Orthop Clin North Am 1983; 4:3-19.
3. Cawley MID. Complications of Paget's disease of bone. Gerontology 1983; 29:276-287.
4. Mirra JM, Brien EW, Tehranzadeh J. Paget's disease of bone: review with emphasis on radiologic features, part 1. Skeletal Radiology 1995; 3:163-171.
5. Freemont AJ. Histology of mineralised tissues. In: Topics in molecular and structural biology, 11. In: Hukins DL, ed. Calcified Tissue. London: The MacMillan Press, 1989:21-40.
6. Freemont AJ. Basic bone biology. Int J Exp Pathol 1993; 74:411-416.
7. Collins DH. Pathology of Bone. London: Butterworths, 1966.
8. Milgram JW. Radiological and pathological assessment of the activity of Paget's disease of bone. Clin Orthop 1977; 127:43-54.
9. Schuller A. Dysostosis hypophysaria. Br J Radiol 1962; 31:156-167.
10. Anderson DC, Richardson PC, Freemont AJ, Cantrill JC. Paget's disease and its treatment with intravenous APD. Advances in Endocrinology 1989; 6:156-164.
11. Harvey L, Gray T, Beneton MNC, Canis J. Ultrastructural fea-

tures of the osteoclasts from Paget's disease of bone in relation to a viral aetiology. J Clin Pathol 1982; 35:771-9.

12. Singer FR and Mills BG. Evidence for a viral etiology of Paget's disease of bone. Clin Orthop 1983; 178:245-251.

13. Cartwright EJ, Gordon MT, Freemont AJ, Anderson DC, Sharpe PT. Paramyxovirus and Paget's disease. J Med Virol 1993; 40:133-141.

14. Hoyland JA, Freemont AJ, Sharpe PT. Interleukin-6 (IL-6), IL-6 receptor and IL-6 nuclear factor gene expression in Paget's disease. J Bone Min Res 1994; 9:75-80.

15. Krane SM, Kantrowitz FG, Byrne M, Pinnell SR, Singer FR. Urinary excretion of hydroxylysine and its glycosides as an index of collagen degradation. J Clin Invest 1977; 59:819-827.

16. DeDeuxchaisnes CN, Krane S. Paget's disease of bone: clinical and metabolic observations. Medicine 1964; 43:233-266.

17. Altman RD, Collins B. Musculoskeletal manifestations of Paget's disease of bone. Arth Rheum 1980; 23:1121-1127.

18. Huvos AG. Bone tumours. Chapter 8. Tumours associated with Paget's disease of bone. Vol II. Philadelphia: WB Saunders, 1991:201-222.

CHAPTER 2

THE EPIDEMIOLOGY OF PAGET'S DISEASE; CLUES TO THE CAUSE?

David C. Anderson

INTRODUCTION

It is tempting in the era of molecular biology to believe that this is the discipline that alone will provide all the *answers* to a disease such as Paget's disease of bone. This may indeed be true, but what about the *questions*? Perhaps the technique is so all-embracing that soon questions will not be necessary, that the questions will follow the answers, but I somehow doubt it.

I believe that the questions will come from a combination of astute clinical observation of how the disease behaves in the individual, combined with epidemiology, or the study of how the disease behaves in the species as a whole. It is therefore worthwhile in a book dedicated to molecular aspects of the disease to include something on both, in order to direct our search for the needle (the cause of the disease) to the most promising parts of the haystack that is concealing it.

Before reviewing what is known of the epidemiology of Paget's disease, I shall begin by briefly summarizing some aspects of the normal osteoclast, and how it functions: of the behavior of Paget's

The Molecular Biology of Paget's Disease, edited by Paul T. Sharpe.
© 1996 R.G. Landes Company.

disease in the individual; and of the differences in its clinical manifestations between individuals. These all need to be taken into account, before conclusions are drawn as to the etiological implications of its epidemiology.

THE FUNCTION OF THE NORMAL OSTEOCLAST
IN RELATION TO THAT IN PAGET'S DISEASE

The osteoclast is a truly extraordinary cell; it is a highly specialized multinucleated giant cell, whose size and multinuclearity
is probably of importance for its bone-resorbing function,[1] which
forms at the bone surface by fusion of bone-marrow-derived precursors. Its most sophisticated structural performance occurs in
cortical bone, where osteoclasts are a highly organized driving force
for remodeling. Once the process has started at one place (for example in response to a local stress fracture exposing uncovered
bone), it appears to progress relentlessly, with a cone of osteoclasts
boring out a tunnel of bone orientated longitudinally.[2] This *cutting cone* is followed by a much more leisurely *closing cone* of osteoblasts laying down new bone radially inwards, so as to almost
completely fill up the original space. Behind the cutting cone there
is intense mitotic activity,[2] some of which is the division of osteoblast precursors. However, it is often ignored that there must be
equally active division of osteoclast precursors, in order to maintain a constant replenishment of new osteoclast units to contribute their nuclei and cytoplasm to the existing osteoclasts, and/or
to fuse with one another.

These local processes are also affected by the overall hormonal
milieu, and particularly by the level of circulating parathyroid hormone, which is believed to stimulate the osteoclast indirectly
through an action on specialized ("PT") cells of osteoblast lineage,
but not the mature osteobast.[3] The possible links of Paget's disease to primary and secondary hyperparathyroidism will be discussed later.

BEHAVIOR OF PAGET'S DISEASE IN THE INDIVIDUAL

I shall confine myself to aspects of the disease that need to be
emphasized in order to appreciate the significance of what is known
of the epidemiology. Here we turn of course to communities where

Paget's disease is extremely common. The disease is capable of achieving frequencies in the community of 10% or more. This, for a chronic disease that is probably infectious in origin, is an exceptionally high frequency (comparable to that of HIV infection in parts of Africa!). Its *apparent clinical frequency* is much lower than this, such that even in those communities where Paget's disease is in reality extremely common, it is regarded by the public at large and even by the local doctors, as quite rare. This is for two main reasons: First, the proportion of patients with the disease showing obvious symptoms and signs is low—probably in the region of 5%.[4] Second, the disease affects bone—a tissue of remarkable resilience and durability. It does not immediately destroy its overall structure, but it accelerates enormously its rate of turnover, and in so doing distorts its microscopic and eventually its macroscopic architecture.[5] Other forces such as gravity and other stresses will ultimately and over many years alter bone shape and strength, and so lead to symptoms—but generally at an advanced age, when body functions are widely expected to be falling apart anyway! Old people and their diseases do not command much interest from the pressure groups of the young, so the disease does not attract the attention it deserves.

Paget's is a *focal* disease of bone, not a *generalized* one (as are osteoporosis, rickets or hyperparathyroid bone disease, for example). Bones that are affected are distributed with some suggestion of preference for the right side of pelvis,[6] humerus[7] and possibly femur.[8] If there is such a tendency for laterality it is, however, small. There seems to be no bone in the body that cannot be the site of Paget's disease, although some sites are more common than others. The propensity seems to be to favor the axial skeleton, including the pelvis and sacrum, the vertebral column, the skull and the major limb long bones.[6,7] The distribution of lesions seems to be remarkably similar to that followed by bone-seeking metastases, with the difference that the latter affect ribs much more frequently, while the tibia, a common site for Paget's disease, is almost never the site of metastasis.

Paget's in long bones seems generally to *begin at one or other end*, presumably at the main site of osteoclasis, and then to be spread linearly through bone by an advancing line of diseased osteoclasts.[9] The leading edge of disease is V-shaped, because the progress of affected osteoclasts is evidently more rapid on the endostial· surface, where they can either move across the surface, or eat through porous cancellous bone, rather than having to tunnel through hard bone as their cortical counterparts must do. Detailed histological studies at postmortem indicate that the advancing edge of disease is always characterized by massively increased osteoclasis, from a line of osteoclasts that are greatly increased in both size and number.[10] Behind this leading edge bone turnover is greatly increased, as osteoblastic increases in parallel with osteoclastic activity. In the skull the disease often appears as *osteoporosis circumscripta*, where for a long distance behind the leading edge, the bone has a porotic appearance that elsewhere is only seen at the leading edge.

Paget's disease spreads through bone at a remarkably constant rate of about 1 cm per year.[9] To spread half the length of the femur, therefore, requires in the region of 20 years; its progress through the hard cortical bone of the shaft may well be much slower than this.

The evidence is overwhelmingly in favor of the hypothesis that the disease is primarily a *focal disease of the differentiated osteoclasts.* It does not behave as a neoplasm of the osteoclasts would be expected to do. It does not metastasize. It does not resemble any of the known neoplasms of bone (although it may be complicated by neoplastic change, usually sarcomatous, and sometimes multifocal). Rather, the disease behaves as an unusual infection, perhaps initially of one osteoclast, which alters its properties in such a way that it increases the recruitment and differentiation of other osteoclasts in the neighborhood.

An attractive hypothesis is that a diseased osteoclast containing the causative agent gains local advantage through the production of growth factors which attract osteoclast stem cells, encouraging them to replicate and the daughter cells to fuse with the infected cell(s). There is good reason to suppose that giant cells, including osteoclasts, can break up as well as fuse, allowing the process to spread through bone in all directions from the initial focus.

The focal nature of the disease throughout the skeleton would therefore result from an initial point in time seeding out of infection through blood-born stem cells of osteoclast lineage. A well-studied example of the disease spreading radially out from an initial focus has been described by us in a patient with osteoporosis circumscripta after treatment with pamidronate, where relapse occurred at the leading edge from initially tiny foci, in which affected osteoclasts appeared to have escaped from the effect of the drug into previously unaffected bone.[11]

EPIDEMIOLOGICAL IMPLICATIONS OF THE BEHAVIOR OF THE DISEASE IN THE INDIVIDUAL

It is obvious from the above that, since epidemiology is concerned with the origin and cause of the disease, in the case of today's patients with Paget's disease *the important environmental influences will have taken place many years before present.* From its rate of spread we can infer that for most patients at the time of diagnosis, the disease has been present for 30 to 50 years or more. These influences may be to do with disease susceptibility—either inherited or acquired; or with exposure to an environmental agent, probably an infection, at a single point in time, which seeds out in bone. It is also possible that genetic or acquired factors may influence the chance of a particular infection taking hold in the osteoclast(s). Before considering the epidemiology, it is worth pausing to reflect what these factors might be.

The most obvious factor to influence osteoclast numbers and activity is parathyroid hormone, and the major influence on its activity within a population is the vitamin D and calcium status. Lack of sunlight and a diet poor in calcium (and vitamin D) increases the level of subclinical secondary hyperparathyroidism which would be seasonal in nature and confined to Northern and Southern latitudes.[12] In susceptible individuals it leads to rickets and osteomalacia; calcium deficiency compounds D deficiency by stimulating a rise in PTH, which in turn accelerates wasteful hepatic degradation of 25-hydroxyvitamin D 3.[13]

Numbers of active osteoclasts per unit skeletal mass are highest in children, whose bone growth depends on mineralization of cartilage spicules at the epiphyseal plates, followed by osteoclastic resorption and deposition of primary spongiosa. Furthermore, as periosteal new bone is added to the cortex, endosteal reabsorption is necessary to increase the marrow space. If Paget's disease is due to a single time-point seeding out of a process which makes the osteoclasts vandalize the bone they are supposed to tend, then this process is likely to depend on the number and receptivity of osteoclasts at the time of seeding.

Other things being equal, it is also reasonable to suppose for a random seeding process spreading through the blood stream, that the seeding number (in the case of a virus, the virus load) is also likely to be important. This is likely to be a function of the dose of infectious agent to which the individual was exposed. This in turn is likely to be a function of the *proximity* of the infectee to the infector at the time of infection, and *the route of infection* (for example whether aerosol or direct into the bloodstream). It will also depend upon *the natural degree of infectiveness of the particular agent*—as evidenced for example in the much greater chance of infection for the same volume of blood for Hepatitis B than AIDS (probably in that case 100-fold). It would depend on the immune status of the individual, and prior exposure to a related agent might confer some protection.

In some circumstances, *climate* might well influence the chance of exposure to an infectious dose of an agent coming from an animal, for example a dog. Paramyxoviruses which are implicated as a cause of the disease (chapter 4) are very susceptible to drying and ultraviolet light.[14]

THE EPIDEMIOLOGY OF PAGET'S DISEASE

ARCHEOLOGICAL AND HISTORICAL EVIDENCE

Paget's is a bone disease which leaves permanent, and in its advanced stages, unmistakable changes in bone structure, many of which will be preserved in buried skeletal remains. This extends to the microscopic structure of the bone.

What has been regarded as the most convincing historical example was reported in a male Anglo-Saxon skeleton by Wells and Woodhouse,[15] from one of several hundred skeletons buried at Jarrow Monastery in Durham, and dated to around 950 AD. Although the bone certainly appears to be involved in Paget's disease, virtually the whole skeleton is affected, which is extremely unusual in true Paget's disease (as opposed to the so-called Juvenile Paget's; see ref. 8, p.143). For this reason I am not entirely convinced by the relationship of this case to what we recognize as Paget's disease in modern man.

One of the best studies of large collections of well documented skeletons is that of Price[16] from Saxon and Medieval cemeteries from the environs of Winchester Cathedral. Particularly convincing is a classical pagetic skull dating from late Saxon times (9th to 10th century AD); several other skeletons from the same era showed evidence of the disease while none were found from (later) medieval skeletal remains excavated from Winchester. Aaron and colleagues also present convincing evidence of Paget's disease in two 16th century skeletons from a burial ground in Wells, one of which shows classical Paget's of the os calcis.[17] Kanis also refers to other less convincing cases in a range of isolated bones from antiquity from the USA, Egypt and France.[8]

There seems little doubt therefore that the disease is at least 1,000 years old. This perhaps throws doubt on the theory that canine distemper virus (CDV) might be the cause, since it is suggested that this disease spread from South America across Europe and the rest of the world during the late 18th century.[14] This is not, of course, convincing evidence that epozootics of a milder strain had not previously occurred in canine or other species in earlier times.

The most famous putative sufferer was the composer Ludwig Von Beethoven, whose tinnitus and eventual deafness (and therefore much of the course of his later musical career as a composer rather than a performer) have been attributed quite convincingly to monostotic Paget's disease of the skull.[18] It is quite clear from the three postmortem examinations that were carried out, that his skull was highly abnormal, being increased throughout to a thickness of half an inch, and, at least in the petrous temporal bone, vascular, with the auditory nerves "shrunken and devoid of neurina."

WORLDWIDE DISTRIBUTION OF THE DISEASE

The first extensive autopsy study to look at the incidence of Paget's disease was conducted in Dresden between the two world wars by Schmorl, who found an incidence of about 3% in the elderly[19] in over 3,000 autopsies. A slightly higher figure of 4% was obtained, also in a postmortem study on 650 individuals, in Sheffield, in Northern UK by Collins[10] in the mid-1950s. This was also a valuable study in defining the histology at the leading edge of the disease, referred to above. These studies conclude that the incidence increases with age, being rare below the age of forty, and higher by a factor of 1.5- to 2-fold in men than in women. This does not tally with our personal experience, where the incidence of cases presenting for treatment is virtually identical between the two sexes, but of course clinical figures in the elderly are confounded by the shorter life expectancy in men than women, as well as other possible factors which might differ between the sexes, that could influence the chances of being diagnosed with the disease.

There is extensive evidence from the work of Guyer, Barker, Gardner and others of the incidence and distribution of Paget's disease in the late 1970s, in the United Kingdom, Europe, the USA and Australia, which complements and extends these previous studies.[20-23] These studies have all used the methodology of reviewing plain radiographs of the pelvis, either taken because of skeletal problems, or for other reasons (e.g., as part of an intravenous pyelogram). Considerable care was exercised to minimize the possible confounding effect of selection of patients for X-ray. Nevertheless, it is possible that even apparently irrelevant investigations such as intravenous pyelography, might be done more frequently in individuals with Paget's disease. However, this is unlikely to have significantly influenced the figures in view of the concordance with the earlier postmortem data. There is good evidence that 75-90% of cases of Paget's disease will have evidence in one or more bones in the lumbar spine pelvis, sacrum or proximal femora. Clearly the missed cases are particularly likely to have only one or two bones affected. There is remarkable consistency between experienced radiologists looking at the same X-ray, and small likelihood of confusion of the X-ray appearance with other diseases such as bony metastases. Other methods have included sending questionnaires to radiologists, asking them to state the frequency with which they see Paget's disease.[24]

These studies show that the highest incidence in the world is in towns in the Northwest of England, where in Lancaster the incidence in people over the age of 50 is 8.3% (men 6.5% and women 10.0%).[22] Interestingly, in nearby Preston, where the overall prevalence was 7.5%, the ratio was reversed, with 8.6% of men and 6.3% of women having the disease (a sex difference reflected in that for the UK as a whole). In two towns (Wigan and Warrington) only 10 miles apart, the prevalences were 6.8% and 3.8% respectively. There was a 4-fold variation across mainland Britain, with a figure of 2.3% in Aberdeen. In Ireland, the incidence in Dublin was similar to Aberdeen, with a figure of 0.9% in Galwey. The same studies yielded figures in the region of 2-3% in France, Belgium, Holland and Germany, with a clear decline as you move South and East across Europe; extremely low levels were found in Greece and Sicily (Fig. 2.1). Similarly, the incidence in Norway and Sweden is very low.

In support of the finding of remarkable local variation in incidence, a radiological study by Rosenbaum and Hanson found a variation between 2- and 15-fold in incidence between Providence, Rhode Island (high) and nearby Lexington, Kentucky (low).[25] Some recent studies point to other isolated foci of Paget's disease in Spain; in particular in the regions around Salamanca, as well as in Madrid and Barcelona.[26] There is a well-known focus of Paget's Disease in the North of Italy, centered on the town of Avellino, discussed further below in connection with bone tumors.[27-29]

In the United States, the disease appears to be similar to the middle European incidence, and more frequent in Northern latitudes, and of very similar frequency between Whites and Blacks.[30]

Elsewhere there is good data from the same group, of the incidence in Western Australia. There has long been evidence of the frequent occurrence of Paget's disease in Perth in Western Australia, particularly among people of British stock. Gardner et al[31] were able to determine from the X-ray folders where the individuals had been born. The incidence in native born Australians was around half that seen in the UK itself, with intermediate frequencies in those who had emigrated (although figures were not available as to where in the UK they had originated). This is therefore strong evidence of a continuing residual environmental influence

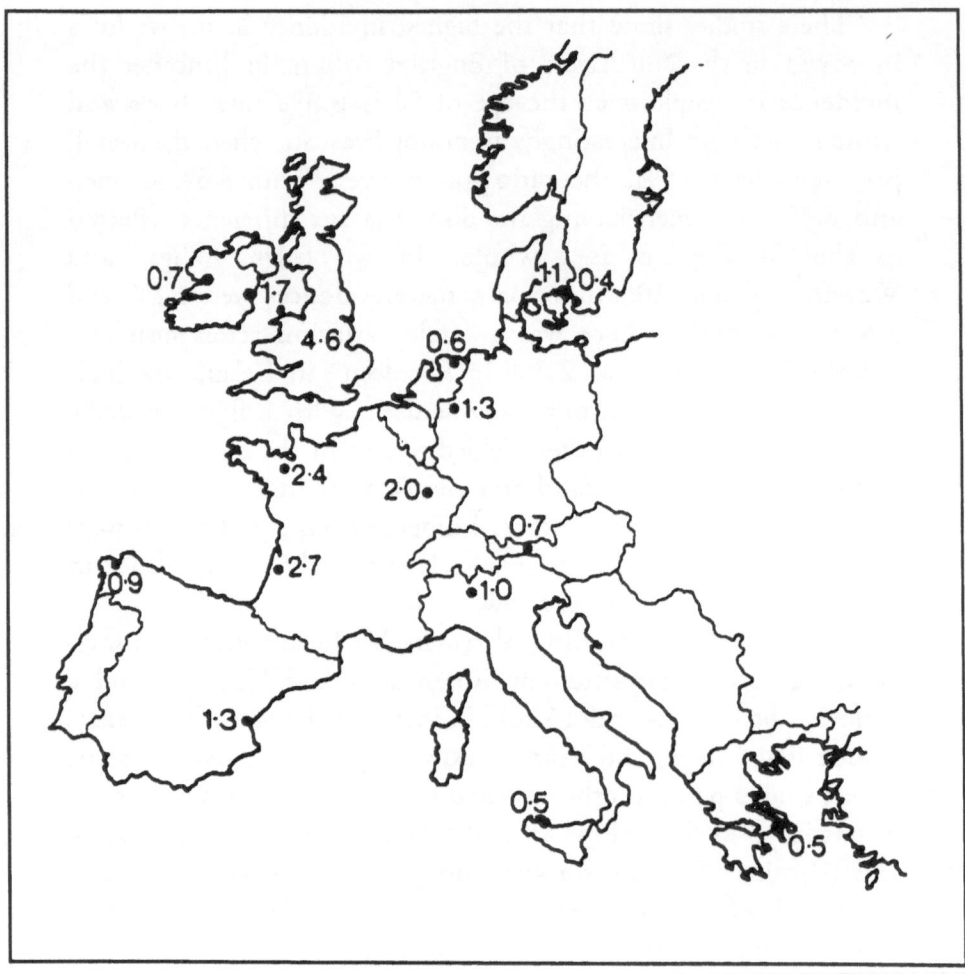

Fig. 2.1. Age and sex-standardized prevalences (percent) of Paget's disease among hospital patients aged 55 years and over in 15 European towns, based on study of plain X-rays. Reprinted with permission from Detheridge FM et al, Br Med J 1982; 285:1005-1008.

from the country of origin, which seems to diminish with the passage of time.

In Eastern Australia the incidence is also high;[32] Posen[33] looked at the countries of origin of 768 patients examined in Sydney between 1975 and 1990, and found that Eastern Europeans were under-represented (3.3% of cases, versus 13.2% of population) while Maltese patients were over-represented (1.8%, versus 0.45%

of population). (As far as I know there are no data on the incidence of the disease in Malta itself.) They also found clear evidence of geographic clustering, with some towns being grossly overrepresented versus their 1921 census figures. Furthermore, in patients born in Sydney, men outnumbered women by a factor of 1.31, while outside the metropolitan area the ratio was 1.43 in favor of women (p < .01).

New Zealand likewise has an incidence of Paget's disease approximating that in the UK.[34] In contrast, the disease is apparently unheard of in the indigenous Maori population.

In Africa, the view until recently was that Paget's disease was rare in Blacks—this is certainly documented for Uganda.[35] However, recent reports suggest that the frequency is quite high in South Africa, and many cases there have been reported in Blacks.[36-38] The same authors report that it is their impression that the disease is more common in South African Whites of British than those of Dutch extraction. Paget's disease has been reported in two Jamaicans of pure African origin.[39]

Another country with a clear focus of Paget's disease is Argentina, where the disease is extremely common in the flat area immediately around Buenos Aires. Mautelen[40] has a very extensive practice of patients with Paget's disease, and has conducted a survey of physicians and radiologists in other South American countries. Elsewhere in South America, the disease appears to be very rare.

In the Middle East it is clear that Paget's disease is quite common in Israel with an incidence in those aged over 55 years of around 1%, comparable to that seen in southern European populations.[41,42] It is, however, only seen among the Jewish population, and particularly in those of European origin. It is extremely rare among Arabs, no Arab patient being reported in a series of 278 cases in Israel.[42]

All the evidence points to an extremely low incidence of Paget's disease in Asia. The disease is almost unheard of in Japan, and although good figures do not exist, it is extremely unusual to see a case in China.[43] In Hong Kong, over 3 years in a busy teaching hospital, with the high probability that all diagnosed cases would be referred to me, I saw only seven definite cases of Paget's disease, and five of these were accumulated cases on the books of other practitioners. One case has been reported from Thailand, in a 44-year old woman of Chinese descent.[44] As far as one can tell, the disease behaves in the same way in oriental as it does in other (non-Asian) individuals.

It is clear that within a particular country foci of Paget's disease exist against a very low background; evidence for the UK, Spain and Argentina has been alluded to above.

FAMILIAL PAGET'S DISEASE

It has long been recognized that there is considerable clustering of Paget's disease within families; specifically, a higher incidence among the first degree relatives.[32] Various forms of inheritance were suggested—including sex-linked recessive and autosomal dominant.[8] One of the most useful and largest family studies was that of Sofaer et al.[45] They analyzed the responses of 407 respondents (68% of those mailed) who belong to the UK National Association for the Relief of Paget's Disease, to a wide range of questions. The prevalence of known Paget's disease was ten times higher among parents and siblings of patients than among the parents and siblings of spouses; the absolute figures were 3.93% and 0.36%, respectively. How this translates into actual relative risk is uncertain, but the chance of having Paget's disease if you come from an area of high incidence and have a first degree relative may be as high as 20% or more, if only one in five cases is diagnosed. Of 56 cases with a family history, 31 came from families where successive generations were affected, and 25 from families where only siblings were affected. Relatives tended to be more alike in regard to age of onset (i.e., diagnosis) than nonfamilial cases. Cases with a positive family history had a significantly earlier age at presentation than those without. In contrast, Kanis[8] reports that in his experience those with a positive family history present at the same age as those without, but that they have more extensive Paget's disease; this seems a strange finding, but may be accounted for by the observation that the "familial" cases have more bones affected on average, but not more extensive disease at each focus. This of course would fit with greater overall exposure of an individual to an infectious agent (viral dose) going in parallel with greater chance of those cohabiting being infected.

An interesting observation in Sofaer's study was that male index cases differed from female index cases in that they had significantly fewer affected male relatives than affected female relatives. Anecdotally I am aware of one family where mother, father and son all have the disease.

A similar high incidence of affected relatives has been found in many other studies, including that of Posen[33] in Australia (22.8% of whose patients reported that they have at least one affected first-degree relative) and our own (22 of 150 (14.6%), compared to 1 out of 300 controls). Siris and colleagues, in a study of cases and spouse controls found a positive family history in 12.3% of cases and 2.1% of controls, and a rate seven times higher in relatives of patients than in relatives of controls.[46] Among relatives, the cumulative risk was greatest when the case had both an early age at diagnosis and bone deformity compared with either one of these features alone. The risk in siblings of a propositus was greater when a parent was affected than when they were not. The findings in this study are consistent with the hypothesis that early age at the time of exposure, and dose of infectious agent, both influence the chance of getting clinically evident Paget's disease many years later.

EVIDENCE FOR DIVERSITY OF THE DISEASE BETWEEN DIFFERENT POPULATIONS

The paper of Jacobs et al[27] reported five patients with Paget's disease and osteoclastomas, of facial bones or vertebrae, in patients with Paget's disease in New York. This is usually a rare bone tumor to be associated with Paget's disease. Three of the patients were related, and all came from Avellino a town in the hills to the East of Naples in Italy, and had emigrated to New Jersey. A fourth patient had multiple family members also affected with Paget's disease. It is unclear whether this variant is due to a different pathogenetic agent, or to a variation in host response. The tumors appear to be very dexamethasone and radio-sensitive, and in one case were multiple. Bhambani and colleagues have likewise reported the case of a 52-year old man who had been born in Avellino with multiple giant cell tumors and aggressive polyostotic Paget's disease.[28] He was not apparently related to any of the cases reported by Jacobs and colleagues. A similar case was reported in the Case records of the Massachusetts General Hospital.[29]

A second related disease, which is certainly not classical Paget's disease, and seems to follow an autosomal dominant inheritance, is the condition found in an extended family in Belfast, named Familial Expansile Osteolysis, discussed in chapter 8.

Wu and colleagues[47] have reported a family of three siblings with Paget's disease, two of whom developed osteogenic sarcomas, and they discuss the possibility that an hereditary component may predispose particular individuals with Paget's disease to develop bone sarcomas.

The question of familial incidence has already been discussed, and is as easily explained on the basis of common exposure of members within a family to an infectious agent, as any inherited tendency. However, there are some instances where the disease spans up to three generations, with multiple members affected. These are clearly suggestive at least of an inherited component.[47]

It should also be mentioned that a similar, but very severe condition is seen sometimes in childhood, and has been labeled juvenile Paget's disease (also called hereditary hyperphosphatasia). This is clearly a different disease.[48]

Otherwise, it appears that Paget's disease, although rare in some races and in some parts of the world, is relatively homogeneous in its clinical manifestations. However, within a given population there are evident differences with some individuals showing evidence of massive new bone formation, in all bones affected, and others showing predominantly osteolytic disease. It seems probable that this response may be related to inherited differences in the osteoblast, for example in its response to vitamin D, as has been suggested to lie at the basis of inherited differences in the tendency to develop osteoporosis.[49] It is entirely possible that such factors affect susceptibility of different individuals.

SEX DIFFERENCES IN THE INCIDENCE OF PAGET'S DISEASE

Most epidemiological studies based on apparently random reviews of pelvic X-rays have indicated a modest (approx. 1.6-fold) male preponderance of Paget's disease.[21,22] Barry,[32] in Australia, concluded that the sex difference was slightly in favor of males (in a proportion of 7 to 6). In contrast, clinical observations[50] or those based on questionnaires to patient groups,[44] identify the problem as being more one of elderly females. It has been suggested that this reflects in part the increasing longevity in females, but also an apparent greater chance of affected females presenting for treatment, perhaps because the disease is more symptomatic, or they have a higher chance of seeing a doctor. No clear HLA associations appear to predispose to Paget's disease.[51]

PERIODICITY IN DATE OF BIRTH

The study of Sofaer and colleagues[44] suggested that when patients were analyzed by year of birth there was a three-year periodicity, consistent with a three-year cycle of infection, to which exposure would have to be within the first two or three years of life. However, this did not achieve statistical significance.

POSSIBLE LINKS TO HIGH EXPOSURE TO DOGS AND/OR OTHER ANIMALS

The extraordinarily high incidence of Paget's disease in the Northwest of England spoke of exceptionally high exposure to an environmental agent many years before present. The evidence, now rather disputed, concerning possible viral inclusion bodies, suggested exposure to and/or excessive susceptibility to, a paramyxovirus which finds sanctuary and flourishes in the osteoclast. We reasoned that the clues must lie in the epidemiology, and it did not seem to fit a common human virus. The next most obvious source was an animal, and clearly domestic pets come high on the list. We therefore questioned our patients in the Manchester area and a control population of diabetics matched for age and sex, about exposure to dogs, cats and budgerigars, and their recalled years of exposure. This study came up with the finding that Paget's patients had a higher recalled exposure to dogs, (twice that of the controls) going back as far as 50 years before present.[52] The study was severely criticized, and others in the UK, notably Barker and colleagues,[53] could not confirm it. However, similar findings were obtained by Ibbotson's group in New Zealand,[54] using a different population of controls, but with the difference that there was an independent association with domestic exposure to cats. A recent report by Kanis' group came up with the interesting observation that patients with Paget's disease were more likely to have a mongrel dog, and less likely to have immunized their dog(s) than were controls.[55]

After our initial study we then went on to study three times as many patients, and matched them each with two controls, each taken from two general practices in Manchester.[56] The findings were similar, though not quite as striking, and showed differences between the two general practice populations as well, but essentially confirmed our initial findings (Fig. 2.2). By this time we were independently pursuing the possibility that canine distemper was the culprit, being a morbillivirus closely related to measles, which affects dogs. These studies are described in chapters 4 and 6.

These findings contrast particularly with those of Siris et al[58] who conducted a questionnaire on more than 400 patients on their register in the United States. They did not find any differences

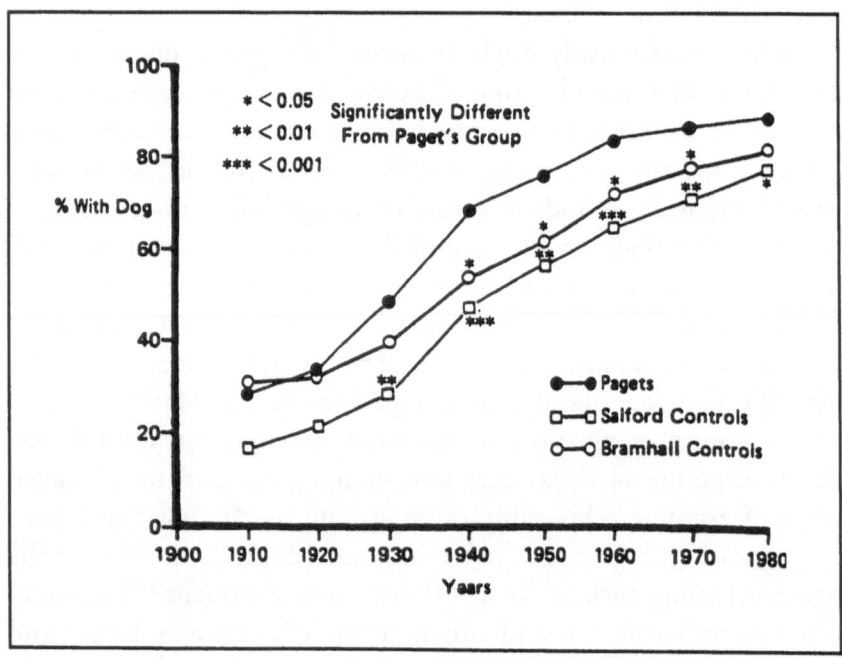

Fig. 2.2. Recalled history of having had a dog in the home, for 150 patients with Paget's disease, and two sets of matched controls from two different general practices in Greater Manchester, given by 10 year cumulative cohorts. Note higher incidence for Paget's patients, most marked for Salford controls. Reproduced with permission from O'Driscoll et al. Bone Miner 1990; 11:209-216.[57]

between patients and controls. However, they used spouses as controls, albeit for other patients; this seems to me to invalidate the study, since spouses generally come from similar backgrounds to their husbands, and indeed have shared much of their life with them! Any differences are therefore likely to be much smaller than they are for nonspouses.

Gordon and colleagues studied a population of my own patients in Salford, for antibodies to canine distemper virus and related this in a subgroup who had undergone bone biopsies to the presence or absence of canine distemper RNA transcripts on in situ hybridization.[59] They found no consistent increase in titer in patients than in control blood (from transfusion donors) but, strangely, found significantly lower levels of antibody in those that were positive for CDV by in situ hybridization. This suggests that there may be some obtuse link to distemper in some cases, but that the immune system is unaware of the continuing infection.

SUGGESTED LINKS TO OTHER ANIMALS

Two other studies have suggested a possible link with animals. Piga et al,[26] in a case control study of Paget's disease sufferers and controls in Spain found a strong association with the consumption of lamb and/or goat without sanitary control (odds ratio 2.18). Lower (and not statistically significant) associations were found with exposure (date not specified) to pets, and a negative association with milk consumption.

A study from Glasgow of antibody levels to murine pneumonia virus (MPV),[60] a virus closely related to respiratory syncytial virus, reported a significantly higher titer of antibodies to MPV in individuals with Paget's disease than in patients with other bone diseases. This clearly warrants further study elsewhere.

POSSIBLE RELATION TO VITAMIN D STATUS
AND CALCIUM INTAKE

Barker and Gardner[20] have suggested that the high incidence of Paget's disease in the Northwest of England may have been related to the high incidence of vitamin D deficiency rickets and osteomalacia in industrial Britain at the turn of the last century, and indeed continuing until fortification of bread with calcium, and childhood supplementation with vitamin D during the second world war. They found from a study of hospital deaths and discharges that progressive cohorts of patients born from 1880 onwards had a progressive decrease in hospital admission and in mortality recorded as being due to Paget's disease (Fig. 2.3) and also in deaths from primary bone tumors in persons aged 55 or more. Most of the latter occur in pagetic bone. The argument for cause and effect is largely circumstantial, and based on the fact that rickets at the turn of the century and Paget's disease in the 1950s and 1960s were both concentrated in industrialized cities in Northern latitudes. This association could also account for the higher incidence of Paget's disease in males, who reportedly had a higher incidence of childhood rickets.[61]

The study of Siris et al[32] found a highly significant association between Paget's disease and aversion to milk products, suggesting a possible association with low calcium intake. A possible link to primary hyperparathyroidism had also been recorded by Posen;[33] patients have a higher recorded incidence of hyperparathyroidism, which might point either to joint association with secondary hyperparathyroidism in earlier life, or a causative association. Again, this is a possible pointer towards increased osteoclastic activity at the time of exposure to the infective agent, as a possible risk factor for the development of Paget's disease.

EVIDENCE FOR A DECLINE IN THE INCIDENCE OF PAGET'S DISEASE

The cohort studies of Barker and Gardner have been discussed above in relation to the possible relevance of childhood vitamin D deficiency. Barker[20,23,62] has presented two lines of evidence to suggest that in the UK Paget's disease might be on the decline—these are first, the mortality and morbidity statistics, and an analysis of where Paget's is recorded on the death certificate. This seems to be declining progressively in cohorts born since the end of the last century. The second line of evidence relates to the declining incidence of primary bone sarcomas, which are usually linked to Paget's in the older age groups. This decline appears to apply both in the UK and in the USA.[62]

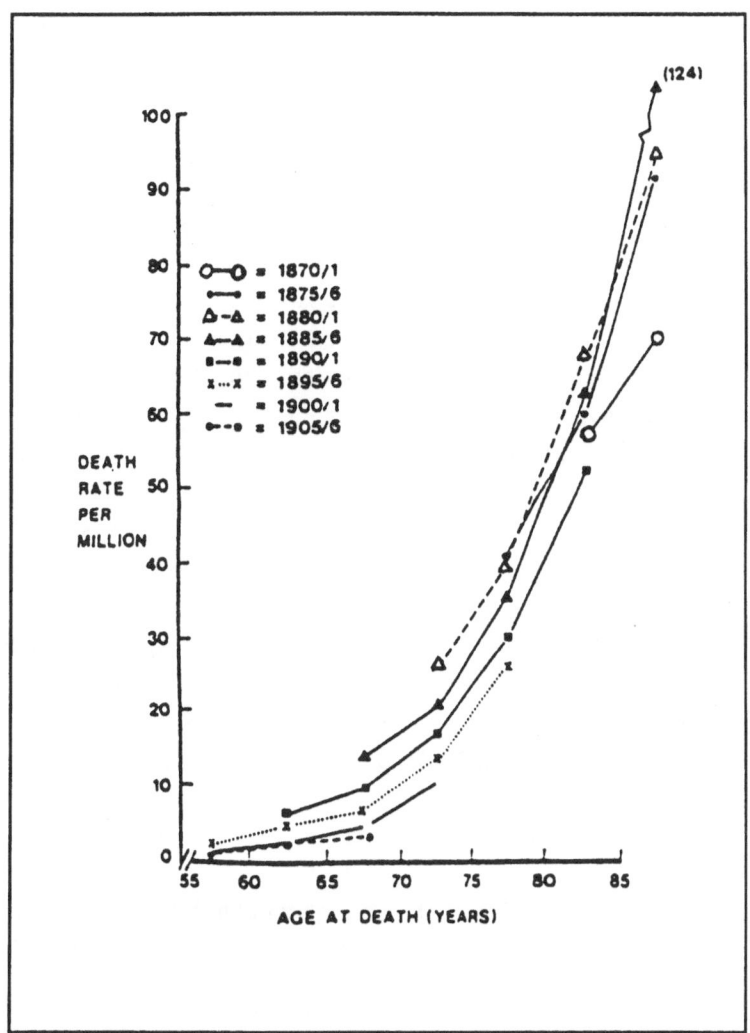

Fig. 2.3. Cohort mortality from Paget's disease in England and Wales in five year periods from 1951-75, by year of birth; note increasing mortality rates for later cohorts, consistent with a decline in incidence of the disease. Reprinted with permission from Barker. Br Med Bull 1984; 40:396-400.

OVERALL CONCLUSIONS, AND THE NEED FOR FURTHER STUDIES!

There is undoubtedly a crying need for a different approach to both the epidemiology and tissue studies of this disease. It seems that the disease behaves as an infection residing in and affecting the fusion and proliferation of osteoclasts. The fact that new foci are so seldom seen speaks against an infection of the blood-born osteoclast precursor. I believe, that at least in the patients in the Northwest of England canine distemper virus is involved. It is surprising that canine distemper virus has only been shown by one group, and it seems intrinsically unlikely that it varies in different locations. However, plenty of examples of synergism between viruses exist (for example the hepatitis B virus and the delta agent; or cytomegalovirus and AIDS). Perhaps CDV is part, but not the whole of the story. Recently a detailed study in the United States by Roodman has provided strong evidence for the presence of another paramyxovirus, measles virus in Paget's disease.[63] In any case it seems highly probable that over many years the virus mutates, and adapts to the particular environment of the osteoclast, and this might include methods of evading the immune system. It is certainly strange that there are not increased levels of antibodies to distemper virus.

There is clearly a need for further epidemiological studies in different populations. Is there really a link with high level of exposure to dogs in early life? If so, is this related wholly to CDV, which is clearly osteotropic, or perhaps to other viruses of canine origin?

REFERENCES
1. Piper K, Boyde A, Jones SJ. The relationship between the number of nuclei of an osteoclast and its resorptive capability in vitro. Anatomy & Embryology 1992; 186:291-299.
2. Jaworski ZFG Hooper C. Study of cell kinetics within evolving Haversian systems. J Anat 1980; 131:91-102.
3. Rouleau MF, Mitchell J, Goltzman D. In vivo distribution of parathyroid hormone receptors in bone: evidence that a predominant osseous target cell is not the mature osteoblast. Endocrinology 1986; 123:187-191.
4. Kanis JA. Clinical features and complications. Chapter 5, In: Pathophysiology and Treatment of Paget's Disease of Bone. Martin Dunitz, 1991.
5. Anderson DC. Paget's Disease. Chapter 12, In: Handbook of Experimental Pharmacology, vol 107. Physiology and Pharmacology of Bone 1993:419-441.
6. Guyer PB, Chamberlain AT, Ackery DM et al. The anatomic distribution of osteitis deformans. Clin Orthop 1981; 156:141.

7. Meunier PJ, Salson C, Mathieu L et al. Skeletal distribution and biochemical parameters of Paget's disease. Clin Orthop Rel Res 1987; 217:37-44.
8. Kanis JA. Pathophysiology and Treatment of Paget's Disease of Bone. Martin Dunitz, 1991.
9. Doyle FH, Banks LM, Pennock JM. Radiologic observations on bone resorption in Paget's disease. Arthritis and Rheumatism 1980; 1205-1214.
10. Collins DH. Paget's disease of bone: incidence and subclinical forms. Lancet 1956; II:51-57.
11. Anderson DC, O'Driscoll JB, Buckler HM et al. Relapse of osteoporosis circumscripta as a lytic ring after treatment of Paget's disease with intravenous 3-amino-1-hydroxypropylidene-1,1-bisphosphonate. Brit J Radiol 1988; 61:996-1001.
12. Stephens WP, Klimiuk PS, Warrington S, Taylor JC, Mawer EB. Seasonal changes in serum 25-hydroxyvitamin D concentration among Asian immigrants. Clin Sci 1982; 63:577-580.
13. Clements MR, Johnson L, Fraser DR. A new mechanism for induced vitamin D deficiency in calcium deprivation. Nature 1987; 325:62-65.
14. Appel MJG, Gillespie JH. Canine Distemper Virus. In: Gard S, Hallauer C, Meyer KF, eds. Virology Monographs. New York: Springer-Verlag, 1972:1-96.
15. Wells C, Woodhouse N. Paget's disease in an Anglo-Saxon. Med Hist 1975; 19:396-400.
16. Price JL. The radiology of excavated Saxon and medieval remains from Winchester. Clin Radiol. 1975; 26:363-370.
17. Aaron JE, Rogers J, Kanis. Paleohistology of Paget's disease in two medieval skeletons. Am J Physiol Anthropol 1992; 89:325-331.
18. Naiken VS. Did Beethoven have Paget's disease of bone? Ann Intern Med 1971; 995-999.
19. Schmorl G. Über Ostitis deformans Paget. Virchows Arch Pathol Anat. 1932; 283:694.
20. Barker DJ, Gardner MJ. Distribution of Paget's disease in England, Wales and Scotland and a possible relationship with vitamin D deficiency in childhood. Br J Prev Soc Med 1974; 28:226-232.
21. Barker DJ, Clough PW, Guyer PB et al. Paget's disease of bone in 14 British towns. Br Med J 1977; 1:1181-1183.
22. Barker DJ, Clough PW, Guyer PB et al. Paget's disease of bone; the Lancashire focus. Br Med J 1980; 280:1105-1107.
23. Barker DJP. The epidemiology of Paget's disease of bone. Br Med Bull 1984; 40:396-400.
24. Detheridge FM, Guyer P B, Barker DJP. European distribution of Paget's disease of bone. Br Med J 1982; 285:1005-1008.

25. Rosenbaum HD, and Hanson DJ. Geographic variation in the prevalence of Paget's disease of bone. Radiology 1969; 92:959-963.

26. Piga AM, Lopez-Abente G, Ibanez AE et al. Risk factors for Paget's disease: a new hypothesis. Int J Epidemiol 1988; 17:198-201.

27. Jacobs TP, Michelsen J, Polay JS et al. Giant cell tumour in Paget's disease of bone. Familial and geographic clustering. Cancer 1979; 44;742-747.

28. Bhambhani M, Lamberty BG, Clements MR, Skingle SJ, Crisp AJ Giant cell tumours in mandible and spine: a rare complication of Paget's disease of bone. Ann Rheum Dis 1992 51:1335-7.

29. Case Records of the Massachusetts General Hospital. Case 1 1986. New Engl J Med (1986) 314:105-113.

30. Siris E. Epidemiological aspects of Paget's disease: family history and relationship to other medical conditions. Sem Arthr Rheum 1994; 23:222-225.

31. Gardner MJ, Guyer PB, Barker DJP. Radiological prevalence of Paget's disease of bone in British migrants to Australia. Brit Med J 1978; I:1655-1657.

32. Barry HC. Paget's Disease of Bone. Livingstone, Edinburgh & London, 1969.

33. Posen S. Paget's disease: current concepts. Aust NZ J Surg 1992; 62:17-23.

34. Raesbeck JC, Goulding A, Campbell DR et al. Radiological prevalence of Paget's disease in Dunedin, New Zealand. Br Med J 1983; 286:1937.

35. Dodge OG. Bone tumours in Uganda Africans. Brit J Cancer 1964; 18:627.

36. Dahniya MH. Paget's disease of bone in Africans. Br J Radiol 1987; 60:113-116.

37. Guyer PB, Chamberlain AT. Paget's disease of bone in South Africa. Clin Radiol 1988; 39:51-52.

38. Pompe van Meerdervoort HF, Richter GG. Paget's disease of bone in South African Blacks. South Afr Med J 1976; 50:1897-1899.

39. Talerman A, Golding JSR, Kirkpatrick D. Bone tumours in Jamaica. J Bone Joint Surgery 1967; 49B: 802-805.

40. Mautelen C, Pumarino H, Blanco et al. Paget's disease: the South American experience. Sem Arthr Rheum 1994; 23:226-227.

41. Dolev E, Samuel R, Foldes J et al. Some epidemiological aspects of Paget's disease of bone in Israel. Semin Arthr Rheum 1994; 23:228.

42. Bloom RA, Libson E, Blank P et al. Prevalence of Paget's disease of bone in hospital patients in Jerusalem: an epidemiologic study. Israel J Med Sci 1985; 21:954-956.

43. Barry H. Orthopaedic surgery in modern China. J Bone Joint Surg 1957; 39B:800-802.

44. Sirikulchayanonta V, Naovaratanophas P, Jesdapatarakul S. Paget's disease of bone—clinico-pathology study of the first case report in Thailand. J Med Assoc Thailand 1992; 75 (suppl 1);136-139.

45. Sofaer JA, Holloway SM, Emery AEH. A family study of Paget's disease of bone. J Epidemiol Comm Health 1983; 37:226-231.

46. Siris E, Ottman R, Flaster JL et al. Familial aggregation of Paget's disease of bone. J Bone Miner Res 1991; 6:495-500.

47. Wu RK, Trumble TE, Ruwe PA. Familial incidence of Paget's disease and secondary osteogenic sarcoma. A report of three cases from a single family. Clin Orthop 1991; 265:306-309.

48. Verrier Jones J, Reed MF. Paget's disease: a family with six cases Brit Med J 1967:4; 90-91.

49. Thompson RC Jr, Gaull GE, Horwitz et al. Hereditary hyperphosphatasia. Studies in 3 siblings. Amer J Med 1969; 47:209-19.

50. Morrison NA, Qi JC, Tokita A et al. Prediction of bone density from Vitamin D receptor alleles. Nature 1994; 367:284-287.

51. Ziegler R, Holz G, Rotzler B et al. Paget's disease of bone in West Germany. Clin Orthop Rel Res 1985; 194:199-204.

52. Gordon MT, Cartwright EJ, Mercer S et al. HLA polymorphisms in Paget's disease of bone. Sem Arthr Rheum 1994; 23:229.

53. O'Driscoll JB, Anderson DC. Past pets and Paget's disease. Lancet 1985; II:919-921.

54. Barker DJP, Detheridge FM. Dogs and Paget's disease. Lancet 1985; II:1245.

55. Holdaway IM, Ibbertson HK, Wattie D et al. Previous pet ownership and Paget's disease. Bone Miner 1990; 8:53-58.

56. Khan P, Brennan G, Newman J, Gray RES, McCloskey E, Kanis J. Paget's disease of bone and unvaccinated dogs. Bone 1996; 19:47.

57. O'Driscoll JB, Buckler HM, Jeacock J et al. Dogs, distemper and osteitis deformans: a further epidemiological study. Bone Miner 1990; 11:209-216.

58. Siris ES, Kelsey JL, Flaster E et al. Paget's disease of bone and previous pet ownership in the United States: dogs exonerated. Int J Epidemiol 1990; 19:455-458.

59. Gordon MT, Bell SC, Mee AP et al. Prevalence of canine distemper antibodies in the pagetic population. J Med Virol 1993; 40:313-317.

60. Pringle CR, Eglin RP. Murine pneumonia virus: seroepidemiological evidence of widespread human infection. J Gen Virol 1986; 67:975-982.

61. Childs B, Cantolino S, Dyke ML. Observations on sex differences in human biology. Bull Johns Hopk Hosp 1962; 110:134.

62. Gardner MJ, Barker DJP. Mortality from malignant tumours of bone and Paget's disease in the United States and in England and

Wales. Int J Epidemiol 1978; 7:121-130.
63. Nuovo MA, Nuovo GJ, MacConnell P et al. In situ analysis of Paget's disease of bone for measles-specific PCR amplified cDNA. Diagn Mol Path 1992; 1:256-265.

CHAPTER 3

PAGETIC OSTEOCLASTS

G. David Roodman, Linda M. McManus
and Anne Demulder

PATHOLOGY AND ULTRASTRUCTURE
OF PAGETIC OSTEOCLASTS

Paget's disease of bone is characterized by intense bone destruction followed by increased bone formation which results in the loss of normal bone architecture. Bone formed in response to the increased osteoclastic bone resorption seen in Paget's disease is of poor quality, leading to such complications as fractures, degenerative arthritis, and neurologic impairment. The resorbed bone is replaced by coarse, dense trabecular bone which is organized in a haphazard fashion. Although the bone production is disorganized and there is very rapid deposition of new bone in Paget's disease, the primary cellular abnormality in patients with Paget's resides in the osteoclast. The osteoblasts appear to be normal but have increased activity in response to the markedly increased bone resorption.

Early pathologic studies of bones from patients with Paget's disease were the first to suggest that Paget's disease is a disease of the osteoclast. Osteoclasts in bone biopsies from pagetic bone can contain up to 100 nuclei per cell in contrast to normal osteoclasts which contain between 3 and 10 nuclei per cell,[1] and pagetic osteoclasts are increased in size compared to normal osteoclasts.

The Molecular Biology of Paget's Disease, edited by Paul T. Sharpe.
© 1996 R.G. Landes Company.

Histomorphometric studies have shown that the number of resorption surfaces are markedly increased in pagetic bone, with the number of resorption surfaces being about ten times greater than that seen normally, and are at least three times greater than that seen in patients with primary hyperparathyroidism.[1] Osteoclast numbers are increased about 10-fold in pagetic bone compared to normal (3.18 ± 2 in pagetic bone versus 0.33 ± 0.23 in nonpagetic areas).[1] Osteoid volume is moderately but significantly increased in pagetic bone compared to normal controls, and osteoid surfaces are markedly increased about four times above normal. Overall, pagetic bone has extensive but thin osteoid borders suggesting that the calcification rate in pagetic bone is slightly faster than the osteoblastic appositional rate. In addition, marrow fibrosis is seen in pagetic bone which decreases in patients who are treated with bisphosphonates or other therapies that decrease osteoclast numbers.

Interestingly, bone not clinically involved with Paget's disease also appears to show evidence of increased bone remodeling in patients with Paget's disease. Approximately 45% of uninvolved bone shows significantly increased trabecular bone resorption and a modest increase in the numbers of osteoclasts. Both Meunier and co-workers[1] and Siris et al[2] have attributed this increased bone remodeling in uninvolved bone from patients with Paget's disease to increased levels of parathyroid hormone, rather than subclinical involvement of these bones with Paget's disease. However, immunoactive parathyroid hormone levels were only increased in 13 of 109 untreated pagetic patients with normal renal function. Thus, at present there is no comprehensive explanation for the increased bone remodeling seen in uninvolved bone from patients with Paget's disease. Potentially, cytokines that increase osteoclast activity may be produced by bone cells in the pagetic lesions, either osteoblasts or the osteoclasts, and released systemically. The cytokines that have been implicated in the pathogenesis of Paget's disease are discussed in chapter 5.

Scanning electron micrographic studies of bone from Paget's patients have confirmed most of the results seen by histomorphometry. Increased amounts of bone in a random array were seen, and the resting surface of bone appeared as an irregular patchwork compared to normal bone.[3] These studies also showed that the resorption surfaces were very abundant and extended irregularly in multiple directions compared to normal bone. These resorption lacunae were often very deep. Boyde et al[4] showed that the volume of resorption lacunae in Paget's bone was more than ten times greater than resorption lacunae in normal bone. In addition, irregularities of the resorption lacunae were also found. These data suggested that the bone-resorbing capacity of pagetic osteoclasts was markedly increased compared to normals. However, Kanis and co-workers (personal communication) have suggested that the bone-resorbing capacity per nucleus in pagetic osteoclasts is actually decreased compared to normal osteoclasts, and that the markedly increased bone resorption results from the greatly increased numbers of osteoclasts present in the pagetic lesion, as well as the massive size of the osteoclast.

Transmission electron microscopy studies have revealed several unique features that differentiate pagetic osteoclasts from normal osteoclasts. Rebel and co-workers[5] were the first to report several cytological abnormalities present in osteoclasts from patients with Paget's disease. The osteoclasts were irregular in shape with multiple extensions and invaginations. The plasma membrane showed an external coat of dense particles consistent with membranes involved in active transport. Occasionally, masses of small electron dense glycogen-like particles were found in the cytoplasm, and the cytoplasm also contained numerous microfilaments of various dimensions. The osteoclasts had typical ruffled borders with a clear zone. The most striking feature of these cells was their nuclei. There were large numbers of nuclei that were polymorphic. Some were smooth and ovoid, while others were badly deformed with multiple indentations. The paranuclear space was dilated and contained clear vesicles. The nuclei had a peripheral distribution of dense chromatin and large nucleoli. In each of these osteoclasts, several nuclei contained microcylindrical inclusion bodies that were filamentous structures of about 150 Å thick. Transverse sections of these inclusions showed a clear center surrounded by dense structures of about 50 Å in diameter. Occasionally, these filaments were closely packed in a paracrystalline array with the interspace reduced to about 50 Å. The filaments were organized in bundles. These nuclear inclusions in Paget's osteoclasts were never seen in normal osteoclasts. Nuclear inclusions were present in all cells. The nuclear inclusions present in pagetic osteoclasts were similar to nuclear inclusions seen in glial cells from patients with progressive multifocal leukoencephalopathy, a disease most probably due to a papava virus.[6,7] In addition, patients who had subacute sclerosing panencephalitis virus infections also showed similar nuclear inclusions.[8] These investigators suggested that Paget's disease may be due to an "external agent." Mills and Singer[9] confirmed that nuclear inclusions were present in osteoclasts of patients with Paget's disease. These investigators studied 18 patients with Paget's disease. Nuclear inclusions similar to those reported by Rebel and co-workers[5] were present in 20-40% of osteoclasts and were present in about one-fourth of the nuclei. Nuclear inclusions were present in all biopsy specimens from patients with Paget's disease. These nuclear inclusions differed from nuclear bodies which are nuclear organelles associated with cellular hyperactivity.[10] They described these nuclear inclusions as most closely resembling the viral nucleocapsids of measles type virus. In addition to the nuclear inclusions, these pagetic osteoclasts also contained cytoplasmic inclusions which were similar to nuclear inclusions from patients with measles virus. Harvey and co-workers[11] found nuclear inclusions in 56-100% of osteoclasts present in Paget's bone biopsy sections viewed by transmission electron microscopy. These nuclear inclusions occupied 15-75% of the nuclear cross-sectional area. Intracytoplasmic inclusions also were seen in 30-40% of osteoclasts in Paget's bone biopsies. Of interest is that treatment of patients with calcitonin or bisphosphonates, although reducing the number of osteoclasts, did not affect the morphology or prevalence of the nuclear inclusions present in these cells. Thus, a viral etiology has been proposed for Paget's disease and is discussed more extensively in chapter 4.

PATHOPHYSIOLOGY OF THE OSTEOCLAST IN PAGET'S DISEASE

As noted above, the morphology of the osteoclast in Paget's disease is abnormal, and the pagetic osteoclasts may have increased bone-resorbing activity. The basis for the abnormalities in osteoclast function in Paget's disease have not been clearly defined. Attempts have been made to develop animal models for Paget's disease by injecting pagetic osteoclasts into nude mice (F.R. Singer, personal communications) but to date none has been successful. Singer and co-workers[12] have isolated pagetic cells from bone biopsy specimens and have cultured these cells in vitro. Although these cells contained nuclear inclusions, they did not release a virus into the media nor could these cells survive for long periods of time. Conditioned media from these cultures could not infect normal bone cells. Of interest is that some mononuclear cells in these cultures expressed paramyxovirus antigens. Basle and co-workers[13] have isolated pagetic osteoclasts and other cells from pagetic bone tissue by treating bone samples from 8 patients with Paget's disease with ethylene diamine tetra-acetic acid. This technique yields variable numbers of osteoclasts, ranging from 10 to 70 osteoclasts per slide. The cells had between 20 to 50 nuclei per cell, and the isolated osteoclasts failed to exclude trypan blue. The cells stained positively for nonspecific esterase that was inhibitable by fluoride, succinic dehydrogenase and acid phosphatase. Thus, these preparations did not yield viable osteoclasts, but were useful for examining the cytochemistry of isolated osteoclasts from Paget's bone.

Recently, Kukita and co-workers[14] have used long-term human bone marrow cultures as a model system for osteoclast formation in patients with Paget's disease. Long-term marrow culture systems have been used as a model system for normal osteoclast formation.[15-21] Multinucleated cells formed in these human marrow cultures express the osteoclast phenotype (multinucleation, contain tartrate-resistant acid phosphatase, cross-react with the 23c6 monoclonal antibody which preferentially binds to osteoclasts, express calcitonin receptors, and form resorption lacunae on calcified matrices).[16,22-24] These workers have cultured bone marrow mononuclear cells from active lesions from patients with Paget's disease. Multinucleated cells differ from normal marrow derived multinucleated cells and are similar to the pagetic cells isolated from bone formed in these cultures (Table 3.1). The multinucleated cells formed in long-term marrow cultures of Paget's disease contained large numbers of nuclei (7 to 90 nuclei per cell), were much larger than multinucleated cells formed in normal marrow cultures, and

Table 3.1. Differences expressed by multinucleated cells (MNC) formed in long-term cultures of Paget's marrow compared to marrow cultures from normals

1. Increased rate of MNC formation
2. Increased sensitivity of MNC to 1,25-dihydroxyvitamin D_3
3. Increased numbers of MNC
4. Increased numbers of nuclei per MNC
5. Increased tartrate-resistant acid phosphatase activity per cell
6. Ultrastructural differences from normal MNC
7. Differences in MNC surface phenotype from normal MNC
8. Only Paget's MNC express paramyxovirus proteins

formed between 10 to 100 times the number of multinucleated cells formed in long-term human marrow cultures. These multinucleated cells contained tartrate-resistant acid phosphatase, cross-reacted with the 23c6 monoclonal antibody, formed resorption lacunae on calcified matrices, and expressed ultrastructural features which were similar to pagetic osteoclasts. They contained pleomorphic nuclei, large amounts of granular material in the cytoplasm, many filamentous structures, and centrioles (Fig. 3.1). In addition, nuclear inclusions were found in these pagetic marrow derived multinucleated cells. However, these nuclear inclusions were not viral inclusions but represented nuclear bodies[10] consistent with the hyperactivity of these cells. The Paget's marrow-derived multinucleated cells had increased levels of tartrate-resistant acid phosphatase per cell, and had an increased sensitivity to 1,25-dihydroxyvitamin D_3. Concentrations as low as 10^{-11} M 1,25 $(OH)_2D_3$ stimulated formation of multinucleated cells formed in pagetic marrow cultures. In contrast, concentrations of 10^{-10} M 1,25 $(OH)_2D_3$ or higher were required for normal marrow multinucleated cell formation. The ultrastructural features of multinucleated cells formed in long-term marrow cultures of marrow derived from Paget's patients are shown in Figs. 3.1 and 3.2. More recently, Mills and her co-workers[25] have tested pagetic marrow-derived multinucleated cells formed in long-term marrow cultures for expression of viral antigens for paramyxoviruses. These investigators found that Paget's marrow derived multinucleated cells ex-

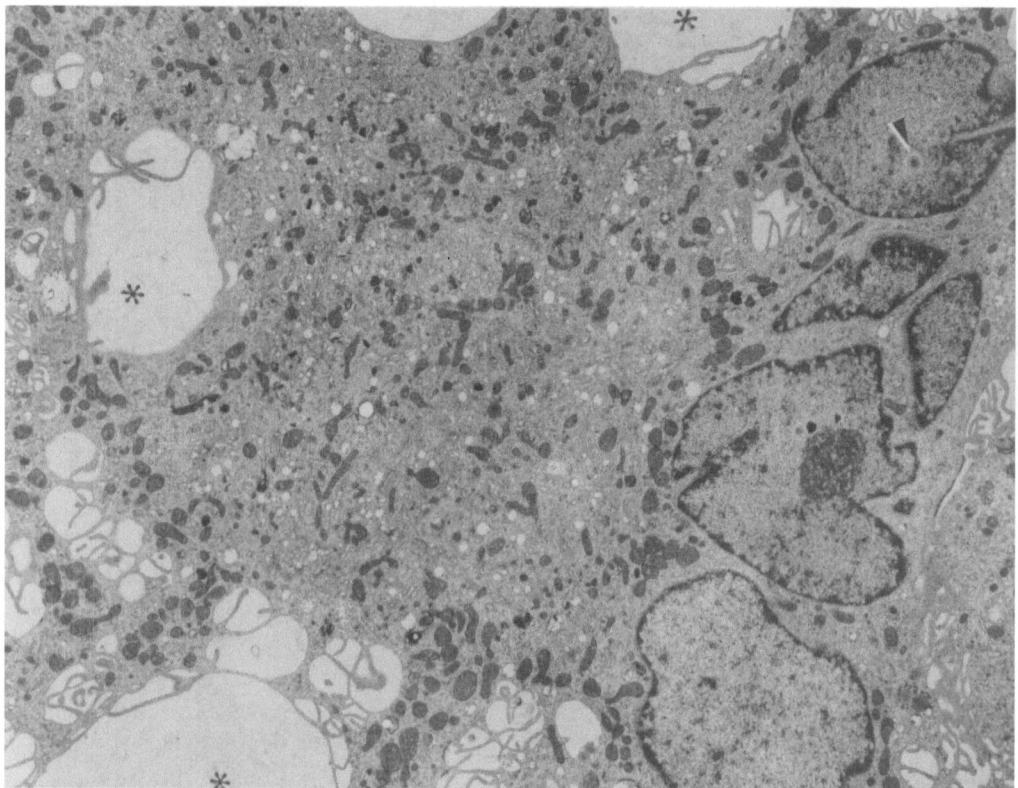

Fig. 3.1. Ultrastructural features of pagetic MNC. Abundant lipid vacuoles are present in this cell which contains prominent intracellular membrane folds surrounding a myelin figure. Open canalicular system (asterisks). Nuclear body (arrowhead). Lead citrate and uranyl acetate. X 8,060.

pressed the nucleocapsid antigens for both respiratory syncytial virus and measles virus, while normal marrow derived multinucleated cells did not express these viral proteins. Thus, the long-term cultures of Paget's marrow may be a model system for examining the factors controlling formation of osteoclasts in Paget's patients. Interestingly, marrow derived from uninvolved bones of Paget's patients also showed similar increases in the number of nuclei per multinucleated cell and in the number of multinucleated cells formed in long-term marrow cultures.[14] These data suggest that there may be a systemic factor present in Paget's patients that enhances osteoclast formation at sites not clinically involved with Paget's disease. These data are similar to results of Meunier and

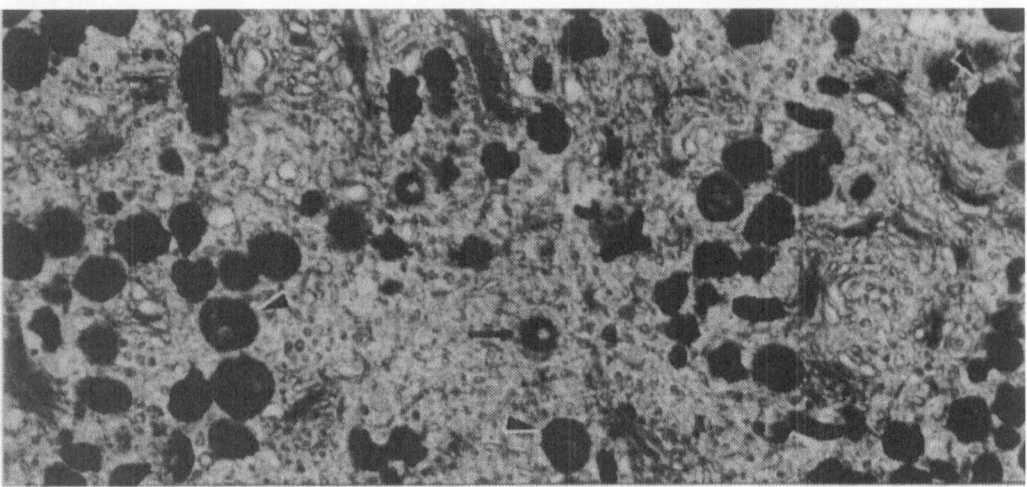

Fig. 3.2. Ultrastructural details of a pagetic MNC. A prominent and extensive Golgi apparatus (arrowheads) is distributed among a variety of intracellular vesicles and granules. Loosely packed juxtanuclear intermediate filaments (arrows) separate many of these organelles. Nucleus (Nu). Lead citrate and uranyl acetate. X 6,240.

co-workers[1] and Siris and co-workers[2] who also found increased os-teoclast activity in bones not clinically involved with Paget's disease.

Recent studies by our group[26] have suggested that interleukin-6 (IL-6) may play a role in the increased osteoclast formation seen in Paget's patients. Bone marrow samples were obtained from pa-tients with Paget's disease and placed in long-term marrow cul-ture. Conditioned media from these long-term marrow cultures were then added to normal marrow cultures and tested for their capacity to increase normal osteoclast-like multinucleated cell for-mation in normal marrow cultures. The Paget's marrow culture conditioned media increased normal multinucleated cell formation. One of the factors present in this conditioned media appeared to be IL-6, because antibodies to IL-6 but not interleukin-1, GM-CSF or TNF-α neutralized the stimulating activity present in the Paget's conditioned media. Further, in situ hybridization studies showed that the multinucleated cells present in the Paget's marrow cul-tures were actively transcribing IL-6 mRNA. When bone marrow plasma from patients with Paget's disease was assayed for IL-6, increased levels were found in 9 of 10 patients while elevated levels of IL-6 were not found in 10 of 10 normal marrow plasma samples. Further, in 19 of 27 patients with Paget's disease, increased levels of IL-6 were found in their peripheral blood plasma while elevated levels of IL-6 were not found in normal marrow plasma from a similar number of controls. These data suggest that IL-6 may be an autocrine/paracrine factor in Paget's disease of bone. Further studies will be required to demonstrate that IL-6 is actively transcribed by osteoclasts isolated from Paget's bone. Suda and co-workers[27] and Lowik et al[28] have recently shown that IL-6 may play a critical role in osteoclast formation and activity. IL-6 increased formation of osteoclasts in bone organ culture systems and also increased osteoclastic bone resorption in newborn mouse calvarial cultures. In addition, Kurihara et al[29] have shown that IL-6 stimulates osteoclast-like multinucleated cell formation in normal marrow cultures and increases the percentage of the cells reacting with the 23c6 monoclonal antibody that preferentially reacts with osteoclasts. The potential role of cytokines in Paget's disease is discussed more extensively in another chapter.

ABNORMALITIES IN OSTEOCLAST PRECURSORS IN PAGETIC MARROW SAMPLES

Demulder and co-workers[30] have used bone marrow culture techniques to study the osteoclast precursors in patients with Paget's disease. The early osteoclast precursors, the CFU-GM, were increased in pagetic marrow samples compared to normals. However, purification of the precursors from marrow stromal cells showed that the absolute numbers of precursors were similar in normal and pagetic marrow samples. Coculture of the pagetic precursors with either normal marrow stromal cells or pagetic marrow stromal cells enhanced the growth of these cells. Furthermore, these pagetic precursors were hyperresponsive to $1,25(OH)_2D_3$ and could form osteoclast-like cells with concentrations of $1,25(OH)_2D_3$ that were 10-fold less than normal cells. These precursors, which can also circulate, expressed measles virus nucleocapsid transcripts when analyzed by PCR. These data suggest that there are abnormalities both in the pagetic osteoclast precursor, as well as the osteoclast, and these culture techniques have been useful to begin to understand the pathophysiology of Paget's disease.

SUMMARY

The osteoclasts in Paget's disease are abnormal both in number and size and show nuclear inclusions as well as cytoplasmic inclusions that suggest a viral etiology for Paget's disease. Use of in vitro model systems for examining osteoclast physiology in Paget's disease has shown that these cells have an increased rate of formation, express high levels of IL-6, IL-6 receptor, and NF-IL-6, as well as most recently, the c-FOS protooncogene.[31] The precursors for these pagetic osteoclasts have several abnormalities including hyperresponsivity to 1,25-dihydroxyvitamin D_3, presence of measles virus nucleocapsid transcripts, and hyperresponsivity to the marrow microenvironment. In addition, the marrow microenvironment is also abnormal in Paget's disease. These data suggest that the initial pathologic event that occurs in Paget's disease affects the osteoclast precursor, as well as the mature osteoclast. The persistence of Paget's disease as a highly localized lesion(s) in patients following diagnosis may be due to the abnormalities in the marrow microenvironment that induce abnormal circulating osteoclast precursors to "home" to the sites of previous disease and induce increased osteoclast formation in these areas. In areas of normal bone, the normal marrow microenvironment does not induce enhanced osteoclast formation, but allows these cells to differentiate toward the monocyte macrophage lineage in the form of mature monocytes. The changes both in the osteoclast and the marrow microenvironment in patients with Paget's disease may account for the persistence of the lesion as a highly localized disease, as well as the enhanced capacity of these cells to form osteoclasts and resorb bone.

REFERENCES

1. Meunier PJ, Coindre JM, Edouard CM, Arlot ME. Bone histomorphometry in Paget's disease. Arth Rheum 1980; 23:1095.
2. Siris ES, Clemens TP, McMahon D, Gordon A, Jacobs TP, Canfield RE. Parathyroid function in Paget's disease of bone. J Bone Min Res 1989; 4:75.
3. Chappard D, Rossi JF, Bataille R, Alexandre C. Osteoclast cytomorphometry demonstrates an abnormal population in B cell malignancies but not in multiple myeloma. Calcif Tissue Int 1991; 48:13-17.

4. Boyde A, Jones J. Estimation of the size of resorption lacunae in mammalian calcified tissues using SEM stereophotogrammetry. In: SEM, Johari O, ed. SEM Inc., O'Hare AMF, II, 1979:393.

5. Rebel A, Malkani K, Basle M, Bregeon CH. Osteoclast ultrastructure in Paget's disease. Calcif Tiss Res 1976; 20:187.

6. Dayan AD. Progressive multifocal leukoencephalopathy. In: Whitty CWM, Hughes JTC, MacCallum FO, eds. Virus Diseases and the Nervous System. Oxford: Blackwell 1973:199.

7. Zu Rhein GM. Association of papova-virions with a human demyelinating disease, progressive multifocal leukoencephalopathy. Progr Med Virol 1969; 11:185.

8. Raine CS, Feldman LA, Sheppard RD, Barbosa LH, Bornstein MB. Subacute sclerosing panencephalitis virus. Lab Invest 1974; 31:42.

9. Mills BG, Singer FR. Nuclear inclusions in Paget's disease of bone. Science 1976; 194:201.

10. Bouteille M, Kalifat SR, Delarue J. Ultrastructural variations of nuclear bodies in human diseases. J Ultrastruct Res 1967; 19:474.

11. Harvey L, Gray T, Beneton MNC, Douglas DL, Kanis JA, Russell RGG. Ultrastructural features of the osteoclasts from Paget's disease of bone in relation to a viral aetiology. J Clin Pathol 1982; 35:771.

12. Mills BG, Singer FR, Weiner LP, Holst PA. Long-term culture of cells from bone affected by Paget's disease. Calcif Tiss Int 1979; 29:79.

13. Basle MF, Mazaud P, Malkani K, Chreitien MF, Moreau MF, Rebel A. Isolation of osteoclasts from pagetic bone tissue, morphometery and cytochemistry on isolated cells. Bone 1988; 9:1.

14. Kukita A, Chenu C, McManus LM, Mundy GR, Roodman GD. Atypical multinucleated cells form in long term marrow cultures from patients with Paget's disease. J Clin Invest 1990; 85:1280.

15. Takahashi N, Yamana H, Yoshiki S, Roodman GD, Mundy GR, Jones SJ, Boyde A, Suda T. Osteoclast-like cell formation and its regulation by osteotropic hormones in mouse bone marrow cultures. Endocrinology 1988; 122:1382.

16. Takahashi N, Kukita T, MacDonald BR, Bird A, Mundy GR, McManus LM Miller M, Boyde A, Jones SJ, Roodman GD. Osteoclast-like cells form in long term human bone marrow but not in peripheral blood cultures. J Clin. Invest 1989; 83:543.

17. MacDonald BR, Takahashi N, McManus LM, Holahan J, Mundy GR, Roodman GD. Formation of multinucleated cells which respond to osteotropic hormones in human long-term marrow cultures. Endocrinology 1987; 120:126.

18. Roodman GD, Ibbotson KJ, MacDonald BR et al. 1,25 $(OH)_2$ Vitamin D_3 causes formation of multinucleated cells with osteo-

clast characteristics in cultures of primate marrow. Proc Natl Acad Sci USA 1985; 82:8213.

19. Chenu C, Pfeilschifter J, Mundy GR, Roodman GD. Transforming growth factor beta inhibits formation of osteoclast-like cells in long-term human marrow cultures. Proc Natl Acad Sci USA 1988; 85:5683.

20. Takahashi N, MacDonald BR, Hon J, Winkler ME, Derynck R, Mundy GR, Roodman GD. Recombinant human transforming growth factor alpha stimulates the formation of osteoclast-like cells in long-term human marrow cultures. J Clin Invest 1986; 78:894.

21. Takahashi N, Mundy GR, Roodman GD. Recombinant human gamma interferon inhibits formation of osteoclast-like cells by inhibiting fusion of their precursors. J Immunol 1986; 137:3544.

22. Kurihara N, Gluck S, Roodman GD. Sequential expression of phenotype markers for osteoclasts during differentiation of precursors for multinucleated cells formed in long term human marrow cultures. Endocrinology 1990; 127: 3215.

23. Kukita T, McManus LM, Miller M, Civin C, Roodman GD. Osteoclast-like cells formed in long-term human bone marrow cultures express a similar surface phenotype as authentic osteoclasts. Lab Invest 1989; 60:532.

24. Kurihara N, Chenu C, Civin CI, Roodman GD. Identification of committed mononuclear precursors for osteoclast-like cells formed in long term marrow cultures. Endocrinology 1990; 126:2741.

25. Mills BG, Frausto A, Singer FR, Ohsaki Y, Demulder A, Roodman GD. Multinucleated cells formed in vitro from Paget's bone marrow express viral antigens. Bone 1994; 15(4):443-448.

26. Roodman GD, Kurihara N, Ohsaki Y, Kukita A, Hosking D, Demulder A, Singer FR. Interleukin-6: A potential autocrine/paracrine factor in Paget's disease of bone. J Clin Invest 1992; 89:46-52.

27. Ishimi Y, Miyaura C, Jin CH, Akatsu T, Abe E, Nakamura Y, Yamaguchi A, Yoshiki S, Matsuda T, Hirano T, Kishimoto T, Suda T. IL-6 is produced by osteoblasts and induces bone resorption. J Immunol 1990; 145:3297.

28. Lowik CWGM, van der Pluijm G, Bloys S, Hoekman K, Bivoet OLM, Aarden LA, Papapoulos SE. Parathyroid hormone (PTH) and PTH-like protein (PLP) stimulate interleukin-6 production by osteogenic cells: A possible role of interleukin-6 in osteoclastogenesis. Biochem Biophys Res Commun 1989; 162:1546.

29. Kurihara N, Bertolini D, Suda T, Akiyama Y, Roodman GD. Interleukin-6 stimulates osteoclast-like multinucleated cell formation in long term human marrow cultures by inducing IL-1 release. J Immunol 1990; 144:426.

30. Demulder A, Takahashi S, Singer FR, Hosking DJ, Roodman GD. Abnormalities in osteoclast precursors and the marrow accessory cells in Paget's disease. Endocrinology 1993; 133(5):1978-1982.
31. Mee AP, Sharpe PT. Dogs, distemper and Paget's disease. Bioessays 1993; 15(12):783-9.

CHAPTER 4

PARAMYXOVIRUSES AND THEIR POSSIBLE ROLE IN PAGET'S DISEASE

Andrew P. Mee

INTRODUCTION

Despite the many studies which have been carried out since Sir James Paget's classical description of the disorder,[1] the exact cause of Paget's disease remains a mystery. Many theories have been proposed, including inflammatory, autoimmune, endocrine and neoplastic etiologies. However, for several years, the consensus of opinion has been that one or more of the paramyxoviruses might be responsible. It is generally thought that, if paramyxoviruses are the cause of the disease, infection occurs in early childhood, and the virus persists in the bone cells to cause disease later in life.

THE PARAMYXOVIRUSES

Paramyxoviruses are enveloped, single stranded RNA viruses, with helical nucleocapsid symmetry (reviewed by Kingsbury).[2] They are very pleomorphic, but usually spherical or globular, although filamentous forms and larger virus particles also occur,[3] reflecting the relative lack of stringency in the budding stage of the virus

The Molecular Biology of Paget's Disease, edited by Paul T. Sharpe.
© 1996 R.G. Landes Company.

assembly process. The helical nucleocapsid is enclosed in a lipid containing envelope which is studded with two types of glycoprotein projections, one with cell fusion (F) activity and another with hemagglutination (HA) and neuraminidase (NA) activities in some viruses of the family. Another important biological property of these viruses is their ability to cause persistent infection upon passage in cell culture.

They are classified into three genera based on antigenic cross reactivities among the members of each genus and on the presence or absence of HA and NA activity:[4]

1. Paramyxoviruses agglutinate mammalian and avian erythrocytes and have NA activity.
2. Morbilliviruses hemagglutinate but lack NA activity.
3. Pneumoviruses exhibit neither HA nor NA activities.

Paramyxoviruses are found in a wide variety of warm blooded vertebrate species, both wild and domesticated (Table 4.1).

The Virion

The viral messenger RNAs are complementary to the virion RNA. Hence, the paramyxoviruses are called "negative-strand" vi-

---

Table 4.1. Genera and some species of paramyxoviruses

Genus	Species	Main Host
Paramyxovirus	Sendai virus (murine parainfluenza virus type 1)	Mouse
	Human parainfluenza viruses types 1-4	Human
	Mumps virus	Human
	Newcastle disease virus	Chicken
	Simian virus 5 (canine parainfluenza virus type 2)	Dog
Morbillivirus	Measles virus	Human
	Canine distemper virus	Dog
	Rinderpest virus	Cow
Pneumovirus	Respiratory syncytial virus	Human
	Bovine respiratory syncytial virus	Cow

Adapted from refs. 2 and 8.

ruses, as viral message is conventionally designated as positive.[1] As is typical of negative-strand RNA viruses, the virion is made up of an internal ribonucleoprotein core (the nucleocapsid) and an outer lipoprotein envelope (Fig. 4.1). The envelope consists of a lipid bilayer, associated with a nonglycosylated protein, and covered with glycoprotein spikes, projecting from the outer surface, which mediate virus attachment and penetration.

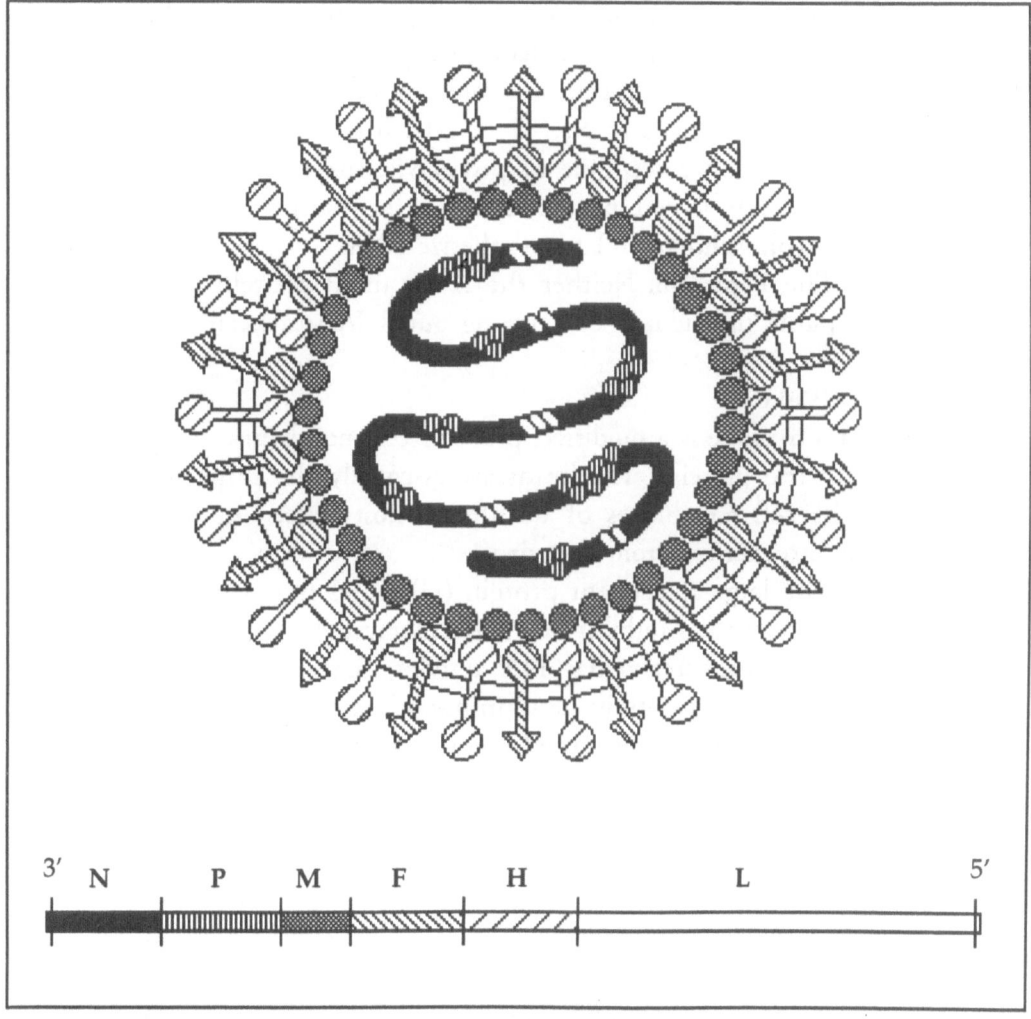

Fig. 4.1. Diagrammatic representation and genetic map of a typical paramyxovirion. Genetic map of a Morbillivirus. N, nucleocapsid protein gene; P, phosphoprotein gene; M, matrix protein gene; F, fusion protein gene; H, hemagglutinin protein gene; L, large protein gene. Adapted from ref. 2.

The Paramyxovirus Genome

The genome consists of a linear molecule of nonsegmented, negative sense, single stranded RNA with molecular weight of about 5×10^6 (approximately 15,000 nucleotides).[1,5] Each genome occupies a single nucleocapsid and the RNA contains a set of six or more genes, linked covalently in tandem (Fig. 4.1).

The Nucleocapsid

The nucleocapsid comprises a single piece of RNA associated with many protein subunits in a rod shaped structure with helical symmetry. There are three nucleocapsid proteins:
1. The major nucleocapsid protein is the structural protein, referred to as N, NP or NC. It is phosphorylated and highly susceptible to proteolytic enzymes.
2. The very large L protein acts together with the smaller P (phosphoprotein) to synthesize RNA.
3. The P protein. Neither the L nor the P protein is capable individually of carrying out RNA synthesis.

The Envelope

The envelope is a modified piece of cell membrane composed of lipids and proteins. The lipids are essentially the same as those found in cell membranes of uninfected host cells, but the three proteins are derived from the virus:
1. The large attachment protein (H) causes hemagglutination in the Morbillivirus and Paramyxovirus genera. In the latter, it also has NA activity, hence it is termed HN (hemagglutinin-neuraminidase). In Pneumoviruses, the corresponding attachment protein lacks both HA and NA activity and has therefore been designated G (glycoprotein).
2. The fusion protein (F), as its name suggests, causes fusion of the virus envelope with the surface membrane of the host cell. It occurs in the form of either a continuous molecule (the inactive precursor known as F_0), or two products of proteolytic cleavage (referred to as F_1 and F_2) which are linked by disulphide bridges to produce the active form.[6,7]

3. The nonglycosylated membrane, or matrix (M), protein plays a role in nucleocapsid-envelope recognition
during viral assembly and it may also participate in the
formation of the envelope.

PARAMYXOVIRUS REPLICATION

Viral replication is confined to the cytoplasm of the host cells
(Fig. 4.2). The replication cycle can be sub-divided into several
different stages:[1,8]

ATTACHMENT AND PENETRATION

The HN, H or G proteins mediate viral attachment and this
is then followed by fusion of the virion envelope with the cell
membrane, allowing delivery of the infective nucleocapsid into the

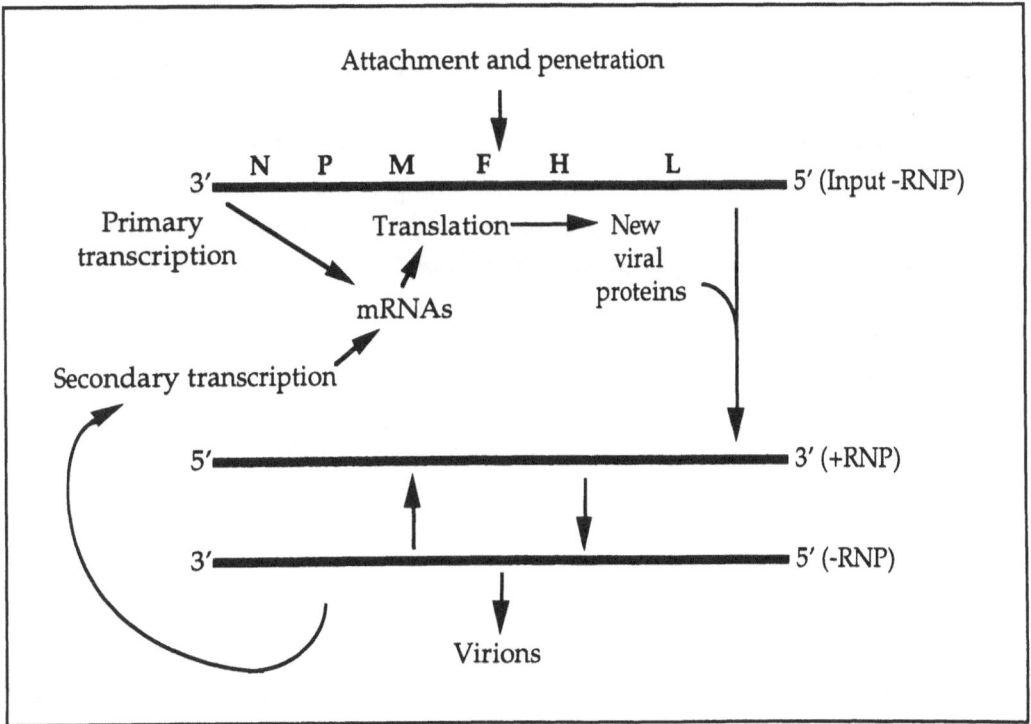

Fig. 4.2. The major steps in paramyxovirus replication. RNP, Ribonucleoprotein; (–), Negative-strand; (+), Positive-strand. Viral replication is confined to the cytoplasm of the host cell, no provirus is formed, and there is no integration into host cell DNA. Adapted from ref. 2.

cell. The fusion process is dependent on the cleavage of F_0 to F_1 and F_2.

TRANSCRIPTION AND TRANSLATION

As the base sequence of the genomic RNA is negative sense, it must first be transcribed before translation can occur. The nucleocapsid acts as a template and, along with the proteins N, P and L, can transcribe the mRNA. Primary transcription takes place under the control of an RNA-dependent RNA polymerase whose activity is due to the L and P proteins.

During infection, each gene is coded for by a separate mRNA species. The mechanism for this is at the transcriptional level, by initiation and termination of RNA synthesis at specific consensus nucleotide sequences which are located at the gene boundaries. There is also transcriptional regulation of mRNA numbers governed by gene order, along with an attenuating effect of the transcriptional regulatory sequences; transcriptional efficiency declines with distance from the 3' end of the genome. This "polarity" is a regulatory mechanism which determines the number of viral proteins, e.g., each nucleocapsid contains approximately 100-fold more N proteins than L proteins; the relative numbers of the other virus proteins reflect their positions in the genome.

Viral mRNAs bind to ribosomes and are translated into protein in the same way as host cell mRNAs.

REPLICATION OF VIRAL RNA

During viral replication, the entire base sequence of the negative-strand RNA is conserved as a single template. The sequences that are not expressed during mRNA synthesis are incorporated into the positive-strand RNA (antigenome) which acts as a template for negative-strand synthesis. Therefore, the RNA synthesis machinery enters an "anti-termination" mode, ignoring the signals at the gene boundaries. The exact mechanism of this switch from mRNA synthesis to replication is not known; however, it is thought to involve the accumulation of a critical concentration of nucleocapsid structure units, which changes the way the RNA polymerase copies its template.

ASSEMBLY AND RELEASE

The initial step in nucleocapsid assembly involves combination of the N structure units with genomic (or, sometimes, antigenomic) RNA. This takes place in the cytoplasm and is followed by addition of the auxiliary nucleocapsid proteins (P and L) to this ribonucleoprotein template.

Assembly of the envelope occurs at the cell surface. Virus glycoproteins (HN/H and F) replace most of the normal cellular proteins in the plasma membrane. This is followed by accumulation of M-protein molecules on the inner surface of the membrane. Nucleocapsids arrive and virions are formed by budding from the cell membrane. As budding takes place, the cell membrane surface becomes studded with viral glycoprotein spikes. Trypsin-like proteases, if present, will activate F_0 molecules present on the surface and this can cause fusion with associated cell membranes. Replicating virus genomes are thus able to spread from cell to cell, largely avoiding circulating anti-viral antibodies. Fusion can take place between a number of cells, leading to the formation of large multinucleated cells (syncytia).

As a consequence of replication, intracytoplasmic inclusion bodies (consisting of N and possibly P, M and L proteins) form. Morbilliviruses also cause the formation of intranuclear inclusion bodies. The precise nature and function of the intranuclear inclusions is unclear, as there is no nuclear component to paramyxoviral replication. However, recent work with canine distemper virus (CDV) has suggested that these inclusions consist of N protein associated with nucleolar derivatives known as nuclear bodies.[9,10]

PARAMYXOVIRUS PERSISTENCE

Viral persistence can be defined as anything ranging from a simple low-grade infection, where infectious virus is continuously produced (such as lymphocytic choriomeningitis in mice and hepatitis B in humans), to infections where the virus genome is present, but infectious virus is not produced, and the persistent virus genome may be defective in some way. This latter type of persistence is different from latent infections, such as herpes virus infection, where reactivation of virus replication can occur.

The ability of paramyxoviruses to persist in vitro has been long established, and it is now becoming clear that they can also persist in vivo. The mechanisms of paramyxovirus persistence are complex, and vary between different viruses of the family. However, several mechanisms are thought to play a general role in the establishment of persistent infections.

MECHANISMS OF VIRAL PERSISTENCE

There are two main mechanisms by which viruses can persist in tissues:

Regulation of cell lysis:

For a virus to persist, there must be a critical number of infected cells. Therefore, the virus must persist without rapidly damaging or destroying the host cells. For viruses that do not normally cause cell lysis, this is easily achieved. However, for those viruses which normally cause cell death, such as the paramyxoviruses, some regulation of this lytic potential must occur. To do this, these viruses can either regulate their gene expression, or generate variants (mutants) that are less lytic, or that interfere with the growth of the "normal" lytic virus. For example, there are temperature sensitive mutants of various paramyxoviruses, including Newcastle disease virus,[11] Sendai virus[12] and measles virus (MV).[13,14] Plaque morphology mutants have also been described, including CDV.[15,16] Defective interfering (DI) viruses spontaneously arise during infections, due to errors of the polymerase complex. These DI viruses are preferentially amplified and interfere with the replication of normal, nondefective, genomes, thereby favoring the establishment of persistent infections.[17] DI paramyxoviruses have been commonly described, including MV[18] and CDV.[19] Subacute sclerosing panencephalitis (SSPE) is a chronic neurological disease resulting from MV persistence in the CNS. Gene-specific hypermutation has been shown in MV isolates from some cases of SSPE.[20]

Evasion of the immune response:

There are two main mechanisms by which the immune system can control viral infections: cytotoxic T lymphocytes, which destroy viral antigens associated with the major histocompatibility complex proteins (i.e., cell-associated virus), and antibodies (often acting with complement), which can recognize both virally infected cells and free virus particles. There are many ways in which viruses can evade these responses.

One suggested mechanism is a reduction in the amount of virally-encoded antigens expressed on the cell surface.[21] This would decrease the risk of an immune response, and is thought to be due to natural selection of cells in which the virus only synthesizes reduced amounts of viral proteins, possibly due to defects in the genome.

Another mechanism of viral persistence is that of antibody-induced antigenic modulation.[22,23] The suggestion is that the actions of the immune system keep the levels of surface viral antigens below a critical point required for cell lysis by continuously removing the viral glycoproteins. That is, antibodies bind and remove the proteins from the cell surface, but there is insufficient antibody to cause cell lysis. The infected cells would still produce other viral proteins, and these could affect the physiological status of the cell. Support for this theory has come from in vitro studies which have shown that the addition of measles-specific antibody, in the absence of complement, leads to a capping and rearrangement of viral proteins.[24,25] These proteins are then shed from the cell surface as antigen-antibody complexes. In addition, the levels of other intracellular viral proteins are also reduced, and this may increase the possibility of a persistent infection. The possibility that this mechanism may also be involved in vivo has been demonstrated. In new-born hamsters, the presence of maternal neutralizing antibody to MV protects against acute encephalitis following intracerebral inoculation of virus, but allows the development of a persistent MV infection in the CNS.[26] Also, in monkeys, it has been shown that a persistent infection is only established in those monkeys with pre-existing anti-MV antibodies, and not in nonimmune animals.[27] It has also been suggested that this effect may result from the presence of cross-reacting antibodies against other viruses, resulting in a persistent infection without any specific antibody response to that virus.[28]

Cell-mediated immunity is also important in the pathogenesis of persistent paramyxovirus infections. It has been shown, for example, that CD8[+] cytotoxic T cells are required to eliminate persistent MV infections in rats.[29] It has also been shown that MV can alter the functions of T cells, and possibly lead to persistence.[30] Patients with SSPE have a history of recovering from an initial infection with MV, and have normal immune responses to other infectious agents.[31,32] It appears, therefore, that these patients develop an impaired T-cell response to MV. The mechanisms of this are unknown, but it has been suggested that an inappropriate T suppresser cell activity might induce tolerance to MV,[33] or that persistent cells are naturally selected by the immune response, as they contain defective genomes in which the mutations are such that the immune response no longer recognizes the target antigens.[34]

Paramyxoviruses can establish persistent infections in sites other than those initially involved in the infection. To do this, they must be transported to these sites of persistence, and this is usually achieved by infection of circulating lymphoid cells. Both MV and CDV can infect and replicate in lymphocytes and macrophages.[35-37] MV has also been isolated from lymphocytes of patients with SSPE,[38] as has Simian virus 5 (SV5).[39] SV5 has also been found in the bone marrow cells of patients with multiple sclerosis (MS).[40] Interestingly, the latter study also found SV5 in 25% of control patients, and found that parainfluenza virus types 1 and 3 were present in the bone marrow cells of 50% of controls and 25% of the multiple sclerosis patients. These results have led to the suggestion that several of the paramyxoviruses may persist in human tissues, and that they might cause widespread, though clinically inapparent, infections.

PARAMYXOVIRUS PERSISTENCE AND DISEASE IN HUMANS

SUBACUTE SCLEROSING PANENCEPHALITIS

SSPE is a slowly progressing, fatal disease of the CNS, affecting children and young adults (reviewed by Swoveland and Johnson).[41] The first thorough description of the disease, which was then termed lethargic encephalitis, reported the presence of intracytoplasmic and intranuclear eosinophilic inclusion bodies, primarily in the grey matter.[42] A similar condition affecting primarily the white matter, called subacute sclerosing leuko-encephalopathy, was also described,[43] and it was 5 years later that the two conditions were recognized as one disease, which was termed SSPE.[44] The inclusions have since been studied electron microscopically and found to resemble paramyxovirus inclusion bodies. Further evidence to support a paramyxoviral etiology came when elevated levels of anti-MV antibodies were detected in the serum and cerebrospinal fluid of SSPE patients, and immunofluorescence studies demonstrated viral antigens in diseased brains.[45] These studies were followed by isolation of MV from affected brain tissue, although, in all cases, cocultivation of brain tissue with MV-susceptible cell lines was required for isolation. Following these initial reports, further isolates have also been described and full length cDNA clones have been isolated and sequenced from SSPE cases (reviewed by Billeter and Cattaneo).[46]

Isolation of viruses by coculture with susceptible cell lines, and sequencing of cDNA clones from MV RNAs of SSPE brains, has allowed the characterization of several different MV isolates from SSPE tissue.[46] Virus isolates from different SSPE cases vary from each other in their gene expression. Mutations due to errors of the viral RNA polymerase have been found in all genes of MV, although not all genes are necessarily affected in any one case. From these studies, it is apparent that up to 2% of nucleotides are mutated during persistence. Of these mutations, approximately 35% result in amino acid changes. Most of the mutations occur in and around the M gene, and it has been suggested that, as M protein is involved in viral assembly, loss of M protein expression could account for lack of viral budding, and thus favor persistence. In some cases of SSPE up to 50% of U residues have been shown to be mutated to C. It has been suggested that these changes result from hypermutation events, rather than accumulation of several mutations of single nucleotides. The exact mechanism of hypermutation is not known, although it has been postulated that it results from aberrant formation of a double-stranded RNA structure involving the MV genome and M mRNA.[47] After unwinding/modification of this structure, biased mutations could occur by introduction of C residues. Whilst these mutations might play a role in persistence in some cases of SSPE, it has been suggested that these events also occur during "normal" MV infections, resulting in defective virions which fail to survive.[46]

AUTOIMMUNE CHRONIC ACTIVE HEPATITIS

As well as in SSPE, MV is also thought to be involved in autoimmune chronic active hepatitis (AICAH). The liver is a target organ for MV and, although jaundice is not a clinical feature of acute MV infection, subclinical hepatitis is thought to be present in up to 80% of affected adults.[48] Anti-measles antibody levels are higher in patients with AICAH than those with natural measles, and can be as high as those seen with SSPE.[49] Furthermore, studies have shown that, since the introduction of measles vaccination programs, the disease is seen much less commonly in younger age groups (i.e., those that have been vaccinated).[50] In situ hybridization studies have revealed the presence of MV within leucocytes of approximately 70% of patients with AICAH.[51] However, several of the control patients also had MV detectable, and this was also the case when the polymerase chain reaction was used.[52] This again suggests that persistent paramyxovirus infections may be more common than previously thought, although these infections are usually not detectable in normal individuals.

MULTIPLE SCLEROSIS

After it was found that MV was the cause of SSPE, a great deal of attention focused on the possible role of paramyxoviruses in other neurological conditions, particularly multiple sclerosis (MS) (reviewed by ter Meulen and Stephenson).[53] Based on epidemiological and serological studies, various paramyxoviruses have been implicated in MS. These include MV, CDV and SV5.[54-59] However, the evidence for CDV and SV5 involvement has been questioned, and attention has focused on MV involvement. More recently, molecular techniques have been used to examine MS tissues. The first study, using solution hybridization, found no evidence of MV,[60] and this was later confirmed by dot blot analysis.[61] However, some evidence has been found for the presence of MV sequences. Haase et al[62] found MV in one out of four MS brains using in situ hybridization, and in a later study found MV in 13 out of 25 MS brains.[63] However, they also found MV in four out of seven control brains. In a similar study, Cosby et al[64] found MV in two out of eight MS patients and in 1 out of 56 control patients. They found no evidence of rubella virus, CDV, or SV5, and concluded that widespread sampling of diseased brains was needed before drawing firm conclusions as to the presence or absence of MV.

The finding of MV in control tissues, particularly in normal brains, might merely reflect the ability of MV to persist without causing any symptoms. It seems likely that, as well as in cases of SSPE, MV can persist in neurological tissues and that this persistence may be responsible for disease only in some cases.

CROHN'S DISEASE

Crohn's disease is an inflammatory bowel disorder that is characterized by granulomatous lesions in the intestinal submucosa.[65] Recently, it has been suggested that Crohn's disease is the result of a persistent MV infection of the mesenteric microvascular endothelium. Supportive evidence for this hypothesis has come from electron microscopy, in situ hybridization and immunohistochemistry.[66] However, control samples of intestine were also found to be positive for MV. The authors concluded that MV is capable of persistently infecting the intestine, and that Crohn's disease may be the result of a granulomatous vasculitis in response to the virus.

As can be seen from these studies, MV (and other paramyxoviruses) can persist in various human tissues. What has still to be proved, however, is that this persistence actually causes the diseases being studied. The possibility still remains that persistence does not cause pathology, and that the virus is only found when diseased tissue (hence active, leading to activation of the virus) is examined.

EVIDENCE FOR A PARAMYXOVIRAL ETIOLOGY OF PAGET'S DISEASE

The first evidence for a possible viral etiology of Paget's disease came from Rebel et al[67] using electron microscopic studies which demonstrated viral nucleocapsid-like inclusion bodies in pagetic osteoclasts (Fig. 4.3). These findings have since been independently confirmed by several groups.[68-70] More recently, these inclusions have been further investigated and have been shown to contain MV and respiratory syncytial virus (RSV) antigens.[71] Although the presence of these inclusions was once regarded as a specific feature of Paget's disease, viral-like inclusion bodies have since been found in several skeletal disorders, including giant cell tumour,[72] pycnodysostosis,[73] osteopetrosis,[74] familial expansile osteolysis[75] and primary oxalosis.[76] Viral-like inclusion bodies are therefore associated with several bone disorders, and there appears to be little correlation between the presence of inclusions and the degree of resorption.

Immunohistochemical studies have further implicated paramyxoviruses, although the evidence is somewhat confusing, as antigens of several paramyxoviruses have been demonstrated in pagetic os-

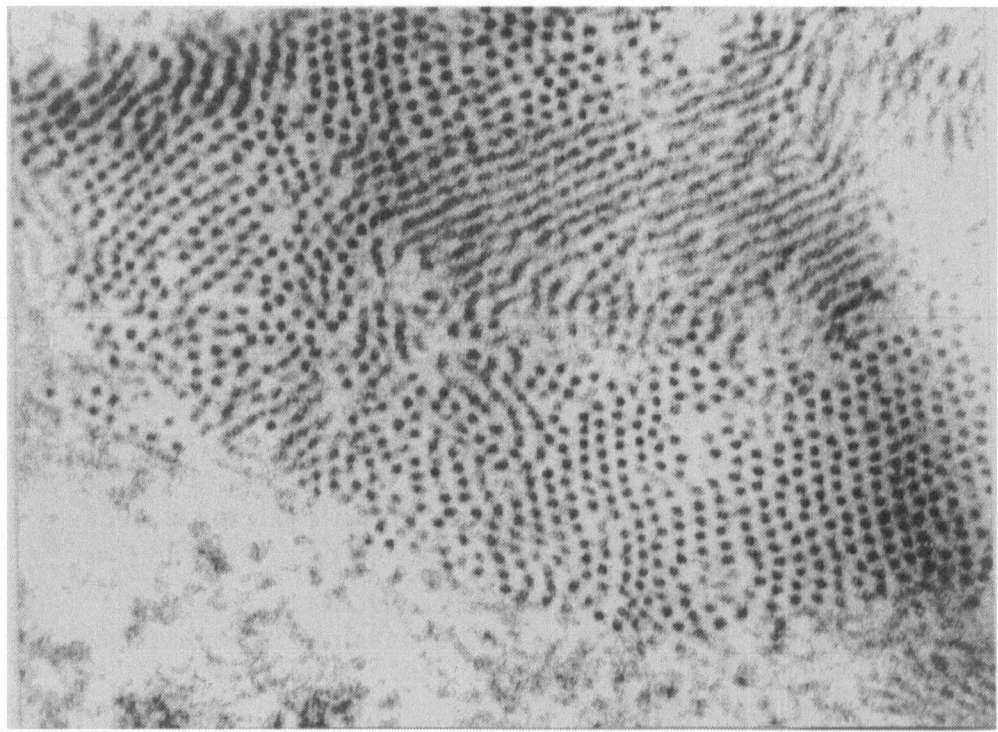

Fig. 4.3. Electron microscopic evidence. High power view of viral inclusion body in a pagetic osteoclast. Note microtubular structure. Mag X 195,000. Figure courtesy of Professor Barbara G Mills, California.

teoclasts, including MV,[77-79] RSV,[78,80,81] parainfluenza virus type 3 and SV5[79] (Fig. 4.4). As with electron microscopy,[71] both MV and RSV have been demonstrated simultaneously in pagetic bone using immunocytochemistry.[82]

Molecular techniques have also provided conflicting results. There has been a study showing positive in situ hybridization with a cDNA probe to MV,[83] although this was not confirmed in later studies using a more specific RNA probe.[84,85] A recent study using reverse transcriptase-polymerase chain reaction (RT-PCR) with degenerate primers, and a further study using primers designed to detect either MV or CDV, both failed to detect any paramyxovirus sequences.[86,87] The technique of in situ-RT-PCR also failed to show any MV sequences.[88] However, RT-PCR has been used to demonstrate MV transcripts in marrow mononuclear osteoclast precursor cells from patients with Paget's disease.[89,90] No MV sequences were detectable in bone samples from pagetic patients, and the authors postulated that failure to detect paramyxoviruses in bone using RT-PCR probably reflected the notoriously difficult problems associated with RNA extraction from this tissue. Sequencing of the MV PCR products obtained from the marrow cells revealed point mutations that resulted in amino acid substitutions (Fig. 4.5a).

These viral findings are, however, difficult to reconcile with the dramatic differences observed in the geographic distribution and prevalence of Paget's disease. Human paramyxoviruses are found in all of the countries in which the disease occurs, suggesting that some other agent might be involved. Paramyxoviruses can infect a wide range of species, including birds and dogs (Table 4.1), and these species do show a more varied geographic distribution.

THE POSSIBLE ROLE OF CANINE DISTEMPER VIRUS IN PAGET'S DISEASE

Recent evidence has suggested that CDV might be involved in Paget's disease. Several epidemiological studies have demonstrated an increased incidence of dog ownership among Paget's patients compared with controls.[91-93] However, these findings have been disputed by others[94,95] (discussed further in chapter 2).

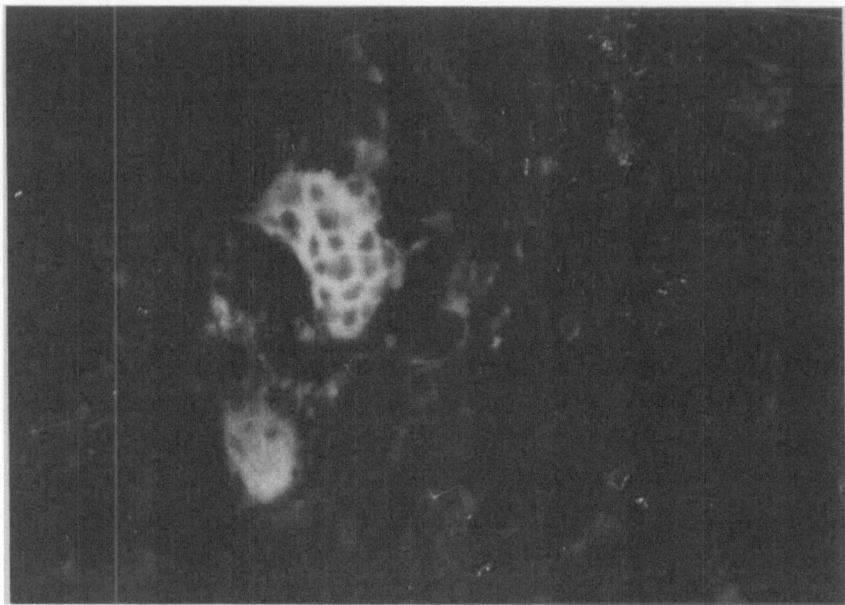

A

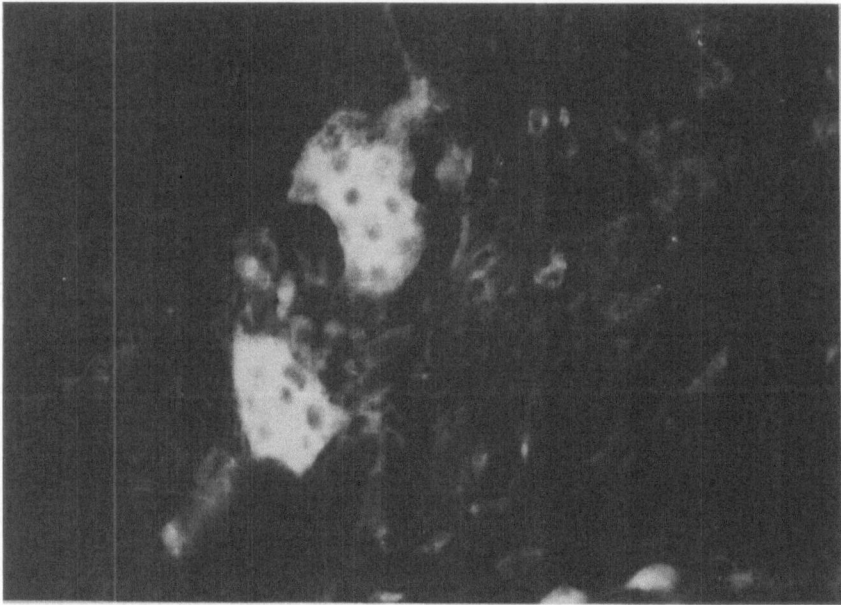

B

Fig. 4.4. Immunocytochemical evidence. Samples from pagetic bone showing the presence of antigens for (A) measles virus and (B) respiratory syncytial virus. Mag X 350. Figures courtesy of Professor Barbara G Mills, California.

(a)

		1351			Thr	Gly	Pro	Ser	1374
EMV	TAC	AGA	GAA	ACC	GGG	CCC	AGC	AGA	

P1   . . .   .   .A. .   . . .
                                                       His 456

P2   . . .   .   G. .   . . .
                                Ala 454

P3   . . .   . .T   . . .   . . .   G. .   . . .
                                                    Gly 457

(b)

		1324	Ser	Asp	Glu	Arg	Leu	Leu	1347
OCDV	TTC	AGT	GAC	GAA	AGG	CTT	CTA	GGG	

Lab   . . .   .   .C.   . . .

P1C1 . . .   .   A. .   . . .   .C.   . . .

P1C2 . . .   .   A. .   . . .   .C.   . . .

P2C1 . . .   .   A. .   . . .   .C.   . . .

P2C2 . . .   .   A. .   . . .   .C.   . . .

P2C3 . . .   C. .   . . .   A. .   . . .   .C.   . . .
            Arg 443   Lys 445   Pro 448

Fig. 4.5. Paramyxovirus sequence data from patients with Paget's disease. (a) Comparison of sequences obtained from pagetic marrow cells[90] with the Edmonston strain of measles virus (EMV) (courtesy of Professor G David Roodman, Texas). (b) Comparison of sequences obtained from pagetic bone[96,98] with the Onderstepoort strain of canine distemper virus (OCDV). Mutated bases and their corresponding amino acid substitutions (if present) are shown. P, patient; C, clone; Lab, laboratory cDNA sequence.

More convincing evidence implicating CDV in the etiology of Paget's disease has come from molecular studies demonstrating the presence of CDV transcripts in pagetic bone cells. Positive in situ hybridization to CDV mRNA has been shown in both osteoblasts and osteoclasts from up to 65% of Paget's patients, using probes to the N, F and P genes of CDV[84,85] (Fig. 4.6). No hybridization was seen with any of the probes to genomic RNA of CDV. This

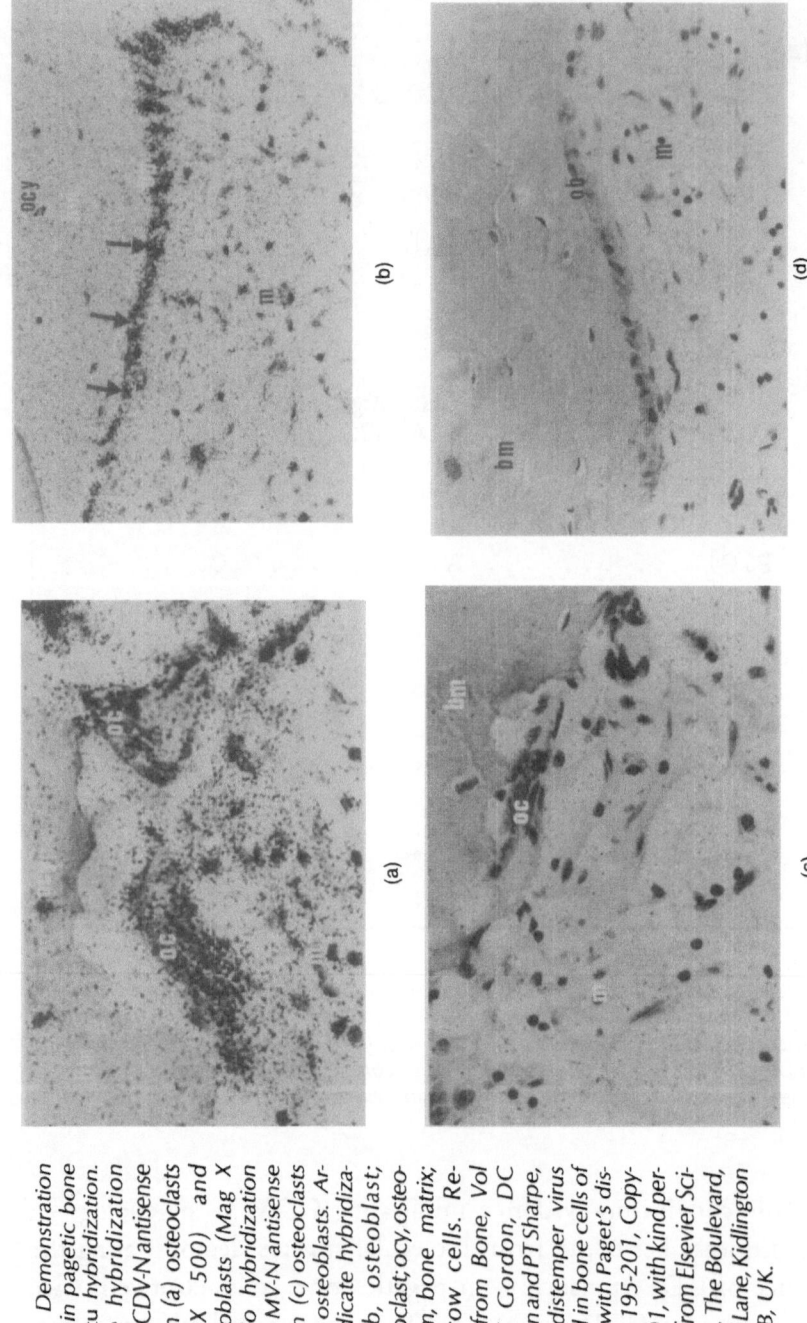

Fig. 4.6. Demonstration of CDV in pagetic bone by in situ hybridization. Positive hybridization with the CDV-N antisense probe in (a) osteoclasts (Mag X 500) and (b) osteoblasts (Mag X 350). No hybridization with the MV-N antisense probe in (c) osteoclasts and (d) osteoblasts. Arrows indicate hybridization. ob, osteoblast; oc, osteoclast; ocy, osteocyte; bm, bone matrix; m, marrow cells. Reprinted from Bone, Vol 12, MT Gordon, DC Anderson and PT Sharpe, Canine distemper virus localized in bone cells of patients with Paget's disease, pp 195-201, Copyright 1991, with kind permission from Elsevier Science Ltd, The Boulevard, Langford Lane, Kidlington OX5 1GB, UK.

finding could be explained by the fact that viral transcription without replication is a known method of viral persistence, or that the virus is in some way replication deficient (see earlier). No patients studied had MV or SV5 viral sequences detectable and only one patient showed positive hybridization for RSV. Further evidence implicating CDV has come from cDNA sequencing following RT-PCR.[96] Using specific primers to the CDV-N gene, RT-PCR showed 8 of 13 Paget's patients had CDV nucleic acids sequestered within their bone cells. This was confirmed by subsequent Southern blotting and probing. Only those patients that were shown to be strongly positive for CDV using in situ hybridization were found to be positive using RT-PCR, supporting the hypothesis of Roodman and his colleagues regarding RNA extraction from bone (see earlier). Both CDV and MV were shown to be present in the bone cells of one patient. The presence of more than one paramyxovirus within the same patient has been documented previously for MV and RSV.[71,82] It was possible to clone and sequence the CDV products from two of the patients, and these sequences revealed approximately 2% base pair changes (5 of 206 bp) in the nucleic acid sequences relative to the Genbank Onderstepoort strain of the CDV-N gene (Fig. 4.5b), similar to the mutation rate seen in MV isolates from cases of SSPE (see earlier). Included in these was a single base pair change (substituting Glu[445] with Lys) compared with the sequence of the cDNA clone present in the laboratory (Fig. 4.5b).[97,98] Hence, it is unlikely that the sequences obtained resulted from contamination with the laboratory clone. Positive PCR results were also found using primers for a region far removed from the laboratory clone. The mutations seen in CDV in Paget's disease could favor persistence at the expense of viral replication, and might explain the lack of genomic RNA found in the in situ hybridization studies.

ANTI-PARAMYXOVIRUS ANTIBODY MEASUREMENTS

Following viral infection, there are usually increased levels of specific anti-viral antibodies; however, measurements of circulating antibodies to several paramyxoviruses have failed to show any significant differences between Paget's patients and controls.[99-102] A recent study investigating 83 Paget's patients also failed to find any significant difference in anti-CDV antibody levels between patients and controls.[103] There was also no apparent change in antibody levels following treatment with 3-amino-hydroxy-propylidene (APD). However, anti-CDV antibody levels were mark-edly raised in several patients and some of the controls. It was also possible with some of the patients to correlate the anti-CDV antibody levels with results from in situ hybridization studies. The levels of anti-CDV antibodies in patients positive for CDV by in situ hybridization were significantly lower when compared with those that were negative for CDV using the same technique. It was postulated that anti-CDV antibodies probably play little or no direct role in the pathogenesis of Paget's disease (which might be expected due to the lack of inflammatory changes seen in Paget's disease), and that failure to clear the virus during an initial infec-tion might favor the sequestration of CDV within the bone cells and eventually lead to Paget's disease.

CANINE DISTEMPER VIRUS IN DOGS

Canine distemper is a highly contagious, acute to subacute dis-ease affecting dogs and other carnivores worldwide (reviewed by Appel).[37] The disease was first thoroughly described in the early 19th century,[104] but it was almost 100 years later before the etio-logic agent was shown to be a filterable virus.[105] The introduction of live, attenuated vaccines has largely controlled the disease, al-though outbreaks still occur in susceptible populations.

Dogs and other carnivores are the natural hosts for CDV, al-though disease signs have been seen in nonhuman primates[106] and seals.[107] Also, a human has been injected subcutaneously with blood obtained from a distemper-infected dog.[108] Although no clinical signs were seen in the man, when his blood was injected intraperi-toneally into two previously uninfected dogs, they developed clini-cal signs of distemper, suggesting that the virus was capable of surviving and replicating in human tissues.

Approximately seven days following natural infection, acutely affected dogs shed virus in all body secretions and excretions, whether they show clinical signs or not.[37] Transmission occurs primarily by aerosol, directly from dog to dog, and the risk of infection is therefore increased in densely populated areas. The virus first replicates in the respiratory tract lymphatic tissues, and then spreads via circulating cells to all other lymphatic tissues. Depending on virus strain, approximately 7-14 days postinfection, dogs either successfully mount a humoral and cellular immune response and recover, or they die from acute or subacute disease, or become persistently infected. In dogs that fail to recover, the virus spreads, via circulating lymphocytes and macrophages, to various epithelia, including those of the respiratory, alimentary and urogenital tracts. CDV first appears in the CNS approximately 8-10 days following infection. Susceptible dogs usually die between 2-4 weeks, depending on the virus strain. Some dogs mount a late immune response, and either succumb to a subacute disease (usually encephalitis), or become persistently infected for two to three months. Many of these dogs eventually recover. Acute encephalitis and death is usually associated with grey matter infection, whereas in subacute and chronic cases, demyelination is also seen. Demyelination has been postulated to be due to "bystander effects" rather than direct CDV infection.[37] It is thought that infected macrophages or astrocytes, which are often found in close proximity to demyelinating axons, release cytotoxic enzymes to cause demyelination. It has also been postulated that reactive oxygen species (ROS) may be involved, as CDV has been shown to stimulate the production of ROS in cultures of canine brain cells.[109] Dogs that develop CNS signs usually die or are euthanased, but some do recover, often with residual signs of disease such as persistent tremors. In addition to acute encephalitis in young dogs, CDV can also cause a chronic encephalitis in more mature animals. This disease is not preceded, or accompanied, by systemic signs.[110] Also, old dog encephalitis, a rare, chronic neurological disease has been postulated to be caused by CDV (see below). Hyperkeratosis of the footpads ("hard pad") is often seen with CDV infection[111] and tooth enamel hypoplasia is a common finding in growing dogs.[112]

OLD DOG ENCEPHALITIS

Old dog encephalitis (ODE) is a subacute or chronic, progressive panencephalitis that occurs in mature dogs. When Cordy[113] first described ODE, he also reported attempts to transmit the disease to other dogs. While he suspected that the disease had an infectious origin, he was unable to confirm this by transmission experiments. To this day, no one has been able to successfully transmit encephalitis to other animals using tissues derived from dogs with ODE. However, further evidence has emerged to implicate CDV in the etiopathogenesis of ODE. CDV antigens have been shown in the brains of dogs with ODE,[114] and anti-CDV antibody levels were also shown to be raised in the serum of these dogs. However, attempts to recover virus and to transmit the disease to other animals were unsuccessful, suggesting that the virus was in some way defective. The same group also provided electron microscopic evidence to support their original findings; paramyxovirus-like inclusion bodies were found in the nuclei of nerve cells.[115] More recently, CDV has been isolated from the brain cells of two of six cases of ODE, however, it was not possible to transmit the disease to ferrets.[116] Interestingly, CDV was also isolated from the bladder of one of the ODE cases, suggesting that CDV can persist in other canine tissues.

CDV AND CANINE RHEUMATOID-LIKE ARTHRITIS

Canine rheumatoid-like arthritis (CRA) is a naturally occurring symmetrical, erosive, inflammatory polyarthropathy.[117,118] On the basis of clinical, pathological and radiological similarities, CRA has been proposed as a model of human rheumatoid arthritis (RA).[119,120] The first evidence implicating paramyxoviruses in CRA came from the detection of paramyxovirus-like inclusion bodies in the synovial membranes of dogs with CRA.[121] CDV has been more specifically implicated by the detection, using an enzyme linked immunosorbent assay, of increased levels of anti-CDV antibodies in the synovial fluids of dogs with CRA compared with those of dogs with osteoarthritis (OA).[119] When the immune complexes were examined by Western blot analysis, those from dogs with CRA were found to react with anti-CDV sera, whilst those from dogs with OA did not react. More recently, it has been shown that levels of anti-CDV antibodies in the sera and synovial fluids of dogs with CRA correlate with the respective total levels of immune complexes.[120] Also, both the synovial fluid anti-CDV antibody levels and the synovial fluid immune-complex anti-CDV antibody levels correlated with the synovial polymorphonuclear cell count, a useful indicator of synovial inflammation. These results suggest an association between CDV and ongoing synovial inflammation in dogs with CRA.

CDV has also been implicated in human RA. Similar inclusion bodies to those found in canine synovial cells are found in the circulating lymphocytes of human RA patients,[122] and epidemiologic studies have revealed an association between RA and dog ownership.[123] Also, anti-CDV antibodies and immune complexes containing CDV antigens have been found within the synovial fluids of humans with RA (Bennett D, personal communication).

CDV AND CANINE BONE DISORDERS

Recently, CDV has been shown to infect and actively replicate in bone cells of young naturally-infected dogs.[124] The virus was concentrated near to the growth plates of the affected dogs and was particularly prominent in the osteoclasts (Fig. 4.7). However, unlike in Paget's disease, both genomic and messenger RNA of CDV was detected, suggesting that active replication of the virus was occurring. This finding might reflect differences between acute and chronic disease, or may be due to differences between infection in humans and the natural host. Further studies using immunocytochemistry on bone samples from experimentally infected dogs have shown that CDV antigen is initially detectable in marrow cells between five and seven days postinfection.[125] Between 8 and 27 days, antigen was found in marrow cells, osteoblasts, osteoclasts and osteocytes, and this was accompanied by mild osseous changes (primarily osteoclast necrosis, associated with subsequent persistence of the primary spongiosa). The numbers of immunopositive cells then declined, until they were undetectable by 41 days postinfection.[125]

The specific localization of CDV in the metaphyses was particularly interesting, as the initial lesions of the canine bone disorder metaphyseal osteopathy (MO) are seen in this area. MO is a skeletal disease of unknown etiology, seen in young, fast-growing dogs, usually of the large and giant breeds. Clinical signs are usually seen between three and six months of age, though they can

Fig. 4.7. Demonstration of CDV in canine bone cells by in situ hybridization. (A) Light- and (B) dark-field views of metaphyseal bone from a naturally distemper-infected dog showing positive hybridization with the CDV-N sense probe. Mag X 160.

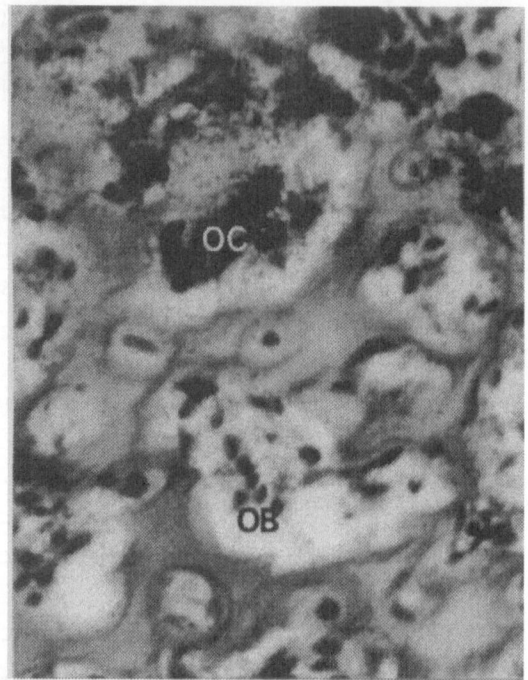

A

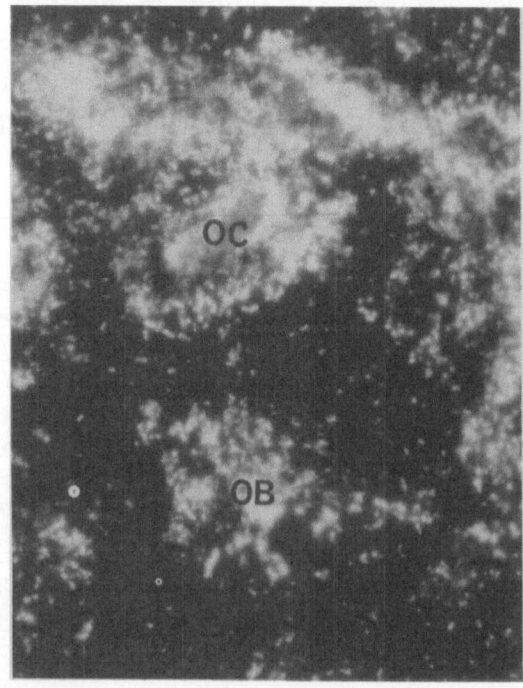

B

occur any time from two months up to two years. There is often a history of some other illness, such as respiratory or gastrointestinal disease, preceding the onset of clinical signs of MO. In the acute stages, these signs consist of fever, anorexia and painful, swollen metaphyses.[126-128] The pain may be so great that affected animals cannot stand. Usually, the faster growing bones, such as the distal radius and ulna, are most severely affected, though many other bones, including the ribs and mandibles, may be involved. Radiographs taken in the acute stages show irregular, alternating radiodense and radiolucent lines parallel to the growth plate.[126-128] Soft tissue swelling is also usually seen. Most dogs recover clinically within a few days, but some go on to develop areas of periosteal and extraperiosteal ossification, resulting in hard, swollen metaphyses. In severe cases, this bone formation may progress to involve almost the whole length of the affected bone,[126-128] giving a gross appearance reminiscent of Paget's disease (Fig. 4.8). These changes can cause permanent deformity, though, if dogs do survive, the excessive bone is usually remodeled and gradually removed. Remissions and relapses sometimes occur and affected dogs can die, though they are more commonly euthanased on humane grounds.

Histologically, changes are seen most prominently in the primary spongiosa, where there is elongation and failure of ossification of the cartilaginous lattice.[127,128] There is also a marked inflammatory infiltrate (neutrophils and lymphocytes) adjacent to the primary spongiosa and extending between the trabeculae, and an increase in the number of osteoclasts. This is accompanied by trabecular necrosis and microfractures. Defective bone formation occurs on the diaphyseal side of the lesion, due to osseous tissue being deposited on fractured and necrotic trabeculae.

Many suggestions have been put forward as to the cause of MO, but none have been proven. The disease was originally thought to be caused by a lack of vitamin C, due to the radiographic similarities to infantile scurvy in humans, and to findings of low serum levels of the vitamin.[126-131] Several authors have claimed successful treatment with ascorbic acid,[132,133] however, others have found that treatment with ascorbic acid had no significant effect on recovery from the disease, as treated and untreated

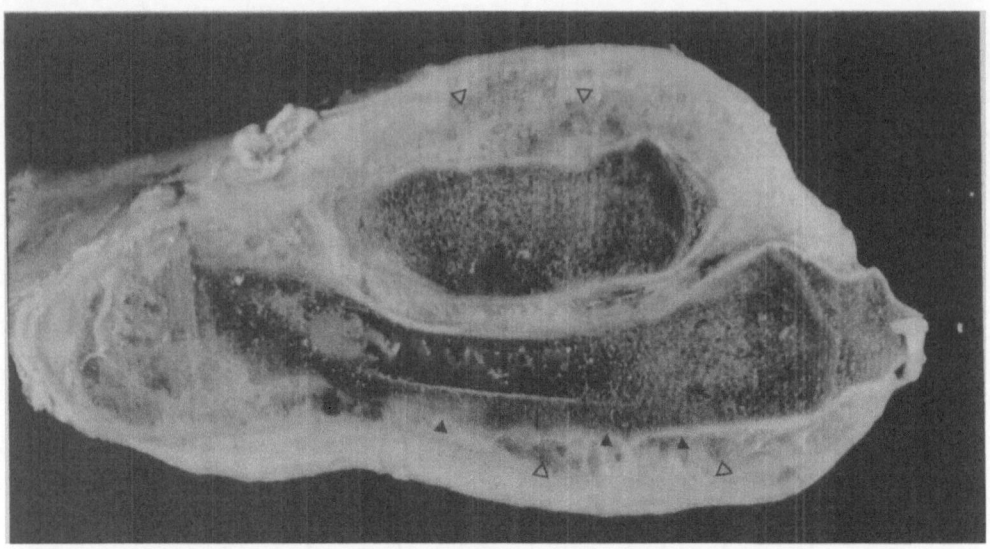

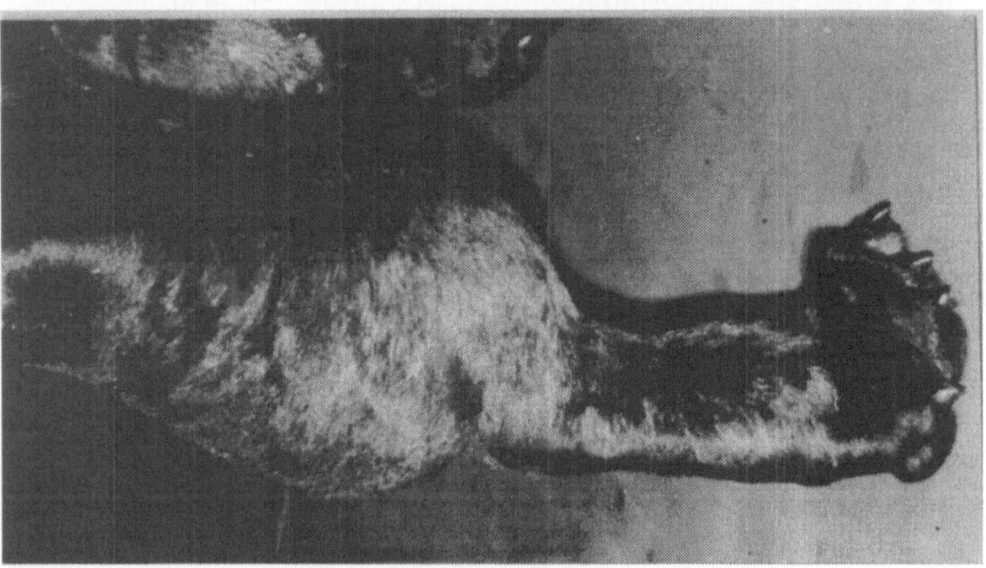

Fig. 4.8. Cross appearance of chronic metaphyseal osteopathy. (A) Distal forelimb of a seven-month-old Great Dane showing swelling and deformity of the antebrachium. (B) Postmortem specimen of the same limb. Note extensive new bone formation (arrows) and calcification and swelling of the soft tissues (open arrows). Reprinted from Bone, Vol 14, AP Mee, MT Gordon, C May, D Bennett, DC Anderson and PT Sharpe, Canine distemper virus transcripts detected in the bone cells of dogs with metaphyseal osteopathy, pp 59-67, Copyright 1993, with kind permission from Elsevier Science Ltd, The Boulevard, Langford Lane, Kidlington OX5 1GB, UK.

dogs had similar rates of recovery.[127,128,134] In Woodard's study of the disease in a litter of Weimaraners,[128] the levels of liver ascorbic acid (which are a more reliable marker of vitamin C status) were measured and found to be within normal limits. One group found that treatment with vitamin C exacerbated the bony lesions,[135] and suggested that vitamin C is contra-indicated in the treatment. Since MO usually spontaneously regresses, and vitamin C levels vary with stress and malnutrition (both of which occur with MO), supposed successful treatments with vitamin C are difficult to assess, and it is now generally accepted that there is some other cause. Another suggestion by some authors, is that overnutrition plays a role in the disease.[136-138] However, the lesions produced experimentally differed both radiographically and histologically from the natural disease and pyrexia was not seen.

Using in situ hybridization, CDV has been demonstrated in metaphyseal bone samples from five dogs with MO.[139] As in Paget's disease,[84,85] only mRNA was detectable, suggesting that little or no viral replication was taking place. The presence of CDV was confirmed by RT-PCR and Southern blotting and probing (Fig. 4.9). CDV was not detectable in bladder and spleen samples from the dogs with MO, suggesting that the localization in bone was specific, and not accompanied by a systemic infection.

EFFECTS OF CDV IN CANINE BONE MARROW CULTURES

Further studies have been carried out to examine the effects of CDV on the formation of osteoclast-like (multinucleated, tartrate resistant acid phosphatase, calcitonin receptor positive) cells from canine marrow mononuclear cells. These have shown that CDV infection (either in vivo or in vitro) significantly increases the number and size (surface area and nuclearity) of osteoclast-like cells formed in the presence of $1\alpha,25$ dihydroxyvitamin D_3 ($1,25(OH)_2$ D_3).[140] Interestingly, the marrow cells became hyper-responsive to $1,25(OH)_2D_3$ following CDV infection. Marrow cells from patients with Paget's disease have also been shown to be hyper-responsive to $1,25(OH)_2D_3$ in long-term cultures,[141] and CFU-GM cells from Paget's patients have been shown to be similarly hyper-responsive to $1,25(OH)_2D_3$.[142] CDV infection also induced the formation of interleukin-6 and c-Fos in the canine marrow cells,[143] supporting the recently proposed molecular model of Paget's disease.[97,144,145]

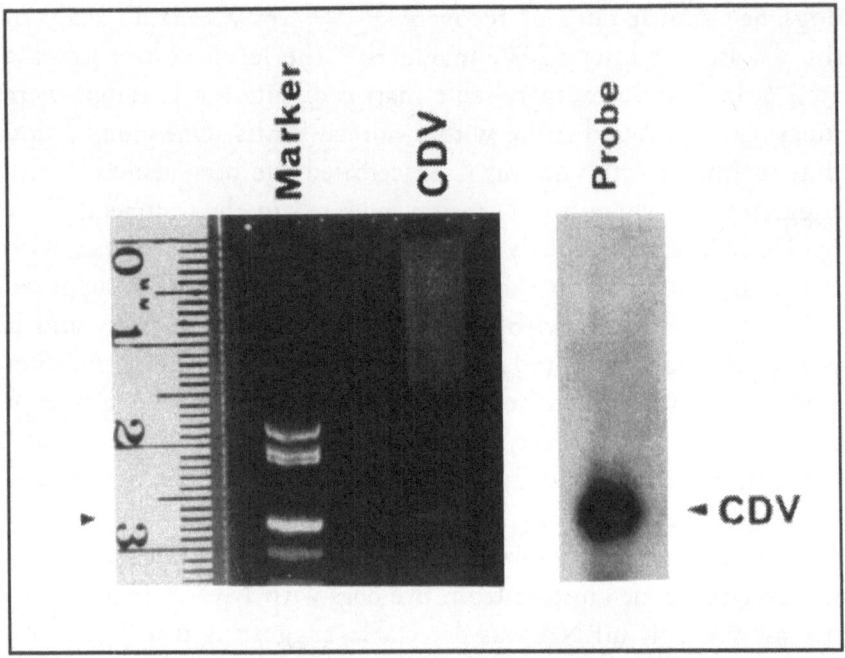

Fig. 4.9. Demonstration of CDV in bone from a dog with metaphyseal osteopathy by RT-PCR and Southern blotting and probing. Band of the expected size for CDV-N (249bp) (arrows) and positive hybridization with the CDV-N cDNA probe in the expected position. Marker–pSP72 cut with Hpa II. Reprinted from Bone, Vol 14, AP Mee, MT Gordon, C May, D Bennett, DC Anderson and PT Sharpe, Canine distemper virus transcripts detected in the bone cells of dogs with metaphyseal osteopathy, pp 59-67, Copyright 1993, with kind permission from Elsevier Science Ltd, The Boulevard, Langford Lane, Kidlington OX5 1GB, UK.

HOW MIGHT PARAMYXOVIRUSES CAUSE PAGET'S DISEASE?

Despite the conflicting evidence regarding the exact identity of the paramyxovirus (or viruses) found in Paget's disease, what is clear is that these viruses can infect human bone cells. Indeed, if paramyxoviruses do cause the disorder, it is possible that any one of several viruses (or a combination) could be responsible. What has yet to be proved; however, is that the virus(es) do actually cause the disease, and are not merely "innocent bystanders"; it is possible that viral persistence in bone cells is a common feature, and causes no pathology. The intense activity of the bone cells seen in Paget's disease could then lead to activation of the virus. Hence, the detection of paramyxoviruses in pagetic bone, by whatever means, might only reflect a consequence of the disease, and not indicate the true cause.

Nevertheless, it is intriguing to investigate the possible mechanisms by which paramyxoviruses might cause Paget's disease. Even if it is assumed that one or more of the paramyxoviruses does cause the disease, several aspects have yet to be explained. How does the virus infect bone cells? How does the virus establish a persistent infection? How does the virus affect the bone cells to cause the bony changes seen in Paget's disease? Why are no "new" lesions found following the initial diagnosis?

Figure 4.10 is a diagrammatic representation of how viruses might infect bone cells and some of the possible consequences of this. Viruses such as MV and CDV are usually spread via the respiratory tract following inhalation, and soon invade the bone marrow. Immunocytochemical and molecular experiments showed the presence of viral proteins or transcripts in marrow cells, osteoblasts, osteocytes and osteoclasts. Hence, it is reasonable to assume that, once within the bone marrow, the virus then has three possible routes of infection; the osteoclast precursors, the osteoblast precursors and/or direct infection of preformed osteoblasts and osteoclasts (Fig 4.10a). Which of these predominates in vivo is uncertain, however, it is likely that all three occur. Infected osteoblasts may progress to become osteocytes, which could act as a possible reservoir of infected cells.

Once the cells are infected, the possible mechanisms by which the virus establishes a persistent infection are those described previously. Many of the infected cells will probably be destroyed by the immune system during the initial infection, however, some must remain, hence the virus must change the way it replicates to evade the immune system. The mutations found in pagetic paramyxovirus sequences are one possible explanation, as is the direct infection of adjacent cells to form syncytia (in this case, osteoclasts), and thus negate the need for the formation of free virus. Evasion of the immune response is a prerequisite for persistence, and would explain the lack of anti-viral antibodies found in serological studies.

Following the establishment of persistence, the infected cells might reasonably be expected to release cytokines and "growth factors" which could then feed back on precursor cells, or directly affect cells within the local environment (Fig 4.10b). The recent

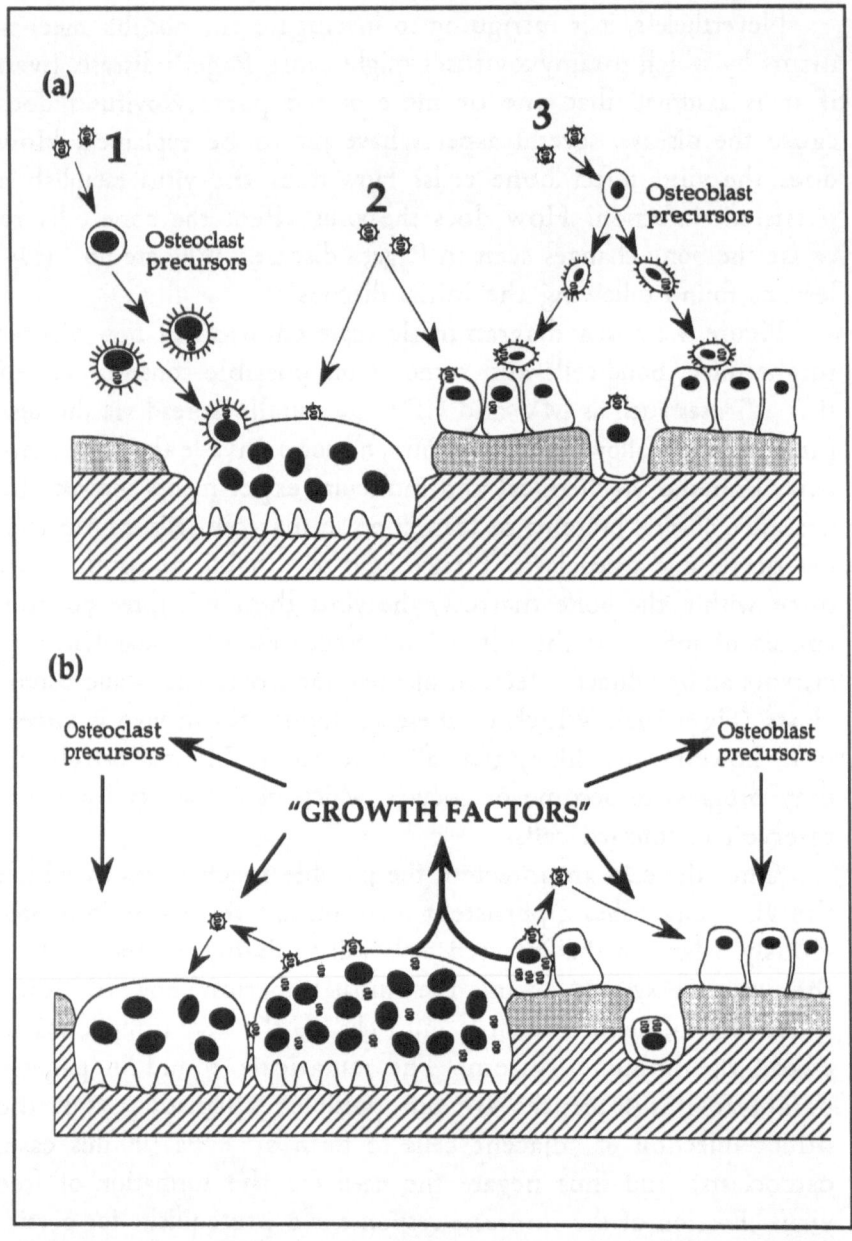

Fig. 4.10. Diagrammatic representation of the possible mechanisms and consequences of paramyxoviral infection of bone. (a) The three possible routes of infection: 1, osteoclast precursors; 2, pre-formed osteoclasts and osteoblasts; 3, osteoblast precursors. (b) Possible consequences of infection. Infected osteoclasts and/or osteoblasts secrete various "growth factors" (such as IL-6) which act on local cells and precursor cells to increase osseous activity. Note that virus could spread by release of free virions or (more likely in order to establish and maintain a persistent infection) by direct cell-to-cell contact. Note also that infected osteocytes could act as reservoirs for future infection of osteoclasts.

biochemical findings in pagetic bone cells provide an interesting and testable hypothesis as to the possible cellular mechanisms of viral infection.

The lack of "new" lesions found once an initial diagnosis has been made would suggest that infection is established at one or more bony sites during the initial viremia, and that dissemination of virus does not occur at any later stage. This means that either spread of virus is inhibited in some way (in order to establish and maintain persistence), or that only local spread occurs. As mentioned previously, the osteocyte could act as a local reservoir to allow the persistence of infection at any particular site. The osteocyte could also be involved in the reactivation of disease/infection that sometimes occurs following treatment with the bisphosphonates.

CONCLUSIONS

The detection of several paramyxoviruses in pagetic bone cells using various techniques would suggest that these viruses are in some way involved in the pathogenesis of the disease. What has yet to be established, however, is any direct causative link. Research into the etiology of Paget's disease is about to enter a new era, as attempts need to be made to establish a link between induction of viral gene expression in bone cells (either using in vitro models as previously described, or transgenic approaches) and a pagetic phenotype, and to examine more closely the biochemical changes that occur in pagetic cells that lead to their altered activities and characteristics. Isolation of viruses from patients may now also be possible since viruses have been demonstrated in marrow osteoclast precursor cells.

ACKNOWLEDGMENTS

Work done by the author was funded by the National Association for the Relief of Paget's Disease and the Salford Paget's Appeal.

I would like to acknowledge the help and advice of various past and present members of the Bone Disease Research Center, especially Margaret Gordon, David Webber, Judith Hoyland, Tony Freemont, Paul Sharpe and David Anderson, and also David

Bennett and Christopher May at the University of Liverpool Small Animal Hospital. I would also like to thank David Roodman for the MV sequence data and Barbara Mills for the EM and immunocytochemistry figures.

REFERENCES

1. Paget J. On a form of chronic inflammation of bones (osteitis deformans). Medico-Chirurg Trans London 1877; 60:37-63.
2. Kingsbury DW. Paramyxoviridae and their replication. In: Fields BN, Knipe DM, eds. Fields Virology, Volume 1, 2nd ed. New York: Raven Press, 1990:945-962.
3. Hosaka Y, Kitano H, Ikeguchi S. Studies on the pleomorphism of HVJ virions. Virol 1966; 29:205-221.
4. Matthews REF. Classification and nomenclature of viruses. Intervirol 1982; 17:104-105.
5. Galinski MS, Wechsler SL. The molecular biology of the Paramyxovirus genus. In: Kingsbury DW, ed. The Paramyxoviruses. New York: Plenum Press, 1991:41-81.
6. Scheid A, Choppin PW. Identification of biological activities of Paramyxovirus glycoproteins. Activation of cell fusion, haemolysis and infectivity by proteolytic cleavage of an inactive precursor protein of Sendai virus. Virol 1974; 57:475-490.
7. Scheid A, Choppin PW. Two disulphide-linked polypeptide chains constitute the active F protein of Paramyxoviruses. Virol 1977; 80:54-66.
8. Choppin PW, Compans RW. Reproduction of Paramyxoviruses. In: Fraenkal-Conrat H, Wagner RR, eds. Comprehensive Virology. New York: Plenum Press, 1975:95-178.
9. Oglesbee M. Intranuclear inclusions in paramyxovirus-induced encephalitis: evidence for altered nuclear body differentiation. Acta Neuropath 1992; 84:407-415.
10. Oglesbee M, Krakowka S. Cellular stress response induces selective intranuclear trafficking and accumulation of morbillivirus major core protein. Lab Invest 1993; 68:109-117.
11. Tsipis JE, Bratt M. Isolation and preliminary characterisation of temperature-sensitive mutants of Newcastle disease virus. J Virol 1976; 18:848-855.
12. Portner A, Marx PA, Kingsbury DW. Isolation and characterisation of Sendai virus temperature-sensitive mutants. J Virol 1974; 13:298-304.
13. Yamazi Y, Black FL. Isolation of temperature-sensitive mutants of measles virus. Med Biol 1972; 84:47-51.
14. Bergholz CM, Kiley MP, Payne FE. Isolation and characterisation

of temperature-sensitive mutants of measles virus. J Virol 1975; 16:192-202.

15. Cosby SL, Lyons C, Fitzgerald SP et al. The isolation of large and small plaque canine distemper viruses which differ in their neurovirulence for hamsters. J Gen Virol 1981; 52:345-353.

16. Cosby SL, Lyons C, Rima BK, Martin SJ. The generation of small-plaque mutants during undiluted passage of canine distemper virus. Intervirol 1985; 23:157-166.

17. Perrault J. Origin and replication of defective interfering particles. Cur Top Microbiol and Immunol 1981; 93:151-207.

18. Morgan EM, Rapp F. Measles virus and its associated diseases. Bacteriol Rev 1977; 41:636-666.

19. Tobler LH, Imagawa DT. Mechanism of persistence with canine distemper virus: Difference between a laboratory strain and an isolate from a dog with chronic neurological disease. Intervirol 1984; 21:77-86.

20. Cattaneo R, Schmid A, Eschle D et al. Biased hypermutation and other genetic changes in defective measles viruses in human brain infections. Cell 1988; 55:255-265.

21. Oldstone MBA, Fujinami RS. Virus persistence and avoidance of immune surveillance: How measles can be induced to persist in cells, escape immune assault and injure tissues. In: Mahy BWJ, Minson AC, Darby GK, eds. Virus Persistence. Cambridge: Cambridge University Press, 1982.

22. Joseph BS, Oldstone MBA. Antibody-induced redistribution of measles virus antigens on the cell surface. J Immunol 1974; 113:1205-1209.

23. Joseph BS, Oldstone MBA. Immunologic injury in measles virus infection. II. Suppression of immune injury through antigenic modulation. J Exp Med 1 975; 142:864-876.

24. Fujinami RS, Oldstone MBA. Antiviral antibody reacting on the plasma membrane alters measles virus expression inside the cell. Nature 1979; 279:935-940.

25. Fujinami RS, Oldstone MBA. Alterations in expression of measles virus polypeptides by antibody: Molecular events in antibody-induced antigenic modulation. J Immunol 1980; 125:78-85.

26. Wear DJ, Rapp F. Latent measles virus infection of the hamster central nervous system. J Immunol 1971; 107:1593-1598.

27. Albrecht P, Burnstein T, Klutch MJ et al. Subacute sclerosing panencephalitis: Experimental infection in primates. Science 1977; 195:64-66.

28. Russell WC, Goswami KKA. Antigenic relationships in the Paramyxoviridae—Implications for persistent infections in the central nervous system. In: Mims C, Cuzner ML, Kelly RE, eds. Vi-

ruses and Demyelinating Diseases. London: Academic Press, 1984:89-99.

29. Maehlen J, Olsson T, Love A et al. Persistence of measles virus in rat brain neurons is promoted by depletion of CD8⁺ T cells. J Neuroimmunol 1989; 21:149-155.

30. McChesney MB, Oldstone MBA. Viruses perturb lymphocyte functions: Selected principles characterising virus-induced immunosuppression. Ann Rev Immunol 1987; 5:279-304.

31. Blaese RM, Hofstrand H. Immunocompetence of patients with SSPE. Arch Neurol 1975; 32:494-495.

32. Sell KW, Ahmed A. Humoral and cellular immune responses in patients with SSPE. Arch Neurol 1975; 32:496.

33. Oldstone MBA. Viral persistence. Cell 1989; 56:517-520.

34. Randall RE, Russell WC. Paramyxovirus persistence. Consequences for host and virus. In: Kingsbury DW, ed. The Paramyxoviruses. New York: Plenum Press, 1991:299-321.

35. Joseph BS, Lampert PW, Oldstone MBA. Replication and persistence of measles virus in defined subpopulations of human leucocytes. J Virol 1975; 16:1638-1649.

36. Sullivan JL, Barry DW, Lucas SJ, Albrecht P. Measles infection of human mononuclear cells. I. Acute infections of peripheral blood lymphocytes and monocytes. J Exp Med 1975; 142:773-784.

37. Appel MJG. Canine distemper virus. In: Appel MJG, ed. Virus Infections of Vertebrates. Amsterdam: Elsevier, 1987:133-159.

38. Horta-Barbosa L, Hamilton R, Witting B et al. Subacute sclerosing panencephalitis: Isolation of suppressed measles virus from lymph node biopsies. Science 1971; 173:840-841.

39. Robbins SJ, Wrzas H, Kline AL et al. Rescue of a cytopathic paramyxovirus from peripheral blood leucocytes in subacute sclerosing panencephalitis. J Infect Dis 1981; 144:396-403.

40. Goswami KKA, Cameron KR, Russell WC et al. Evidence for the persistence of Paramyxoviruses in human bone marrow cells. J Gen Virol 1984; 65:1881-1888.

41. Swoveland PT, Johnson KP. Subacute sclerosing panencephalitis and other paramyxovirus infections. In: McKendall RR, ed. Handbook of Clinical Neurology, Volume 12. Amsterdam: Elsevier, 1989: 417-437.

42. Dawson JR. Cellular inclusions in cerebral lesions of lethargic encephalitis. Am J Path 1933; 9:7-15.

43. Van Bogaert L. Une leuco-encéphalite sclérosante subaigüe. J Neurol, Neurosurg Psych 1945, 8:101-120.

44. Greenfield JG. Encephalitis and encephalomyelitis in England and Wales during the last decade. Brain 1950; 73:141-166.

45. Connolly JH, Allen IV, Hurwitz LJ, Miller JHD. Measles virus

antibody and antigen in subacute sclerosing panencephalitis. Lancet 1967; 1:542-544.

46. Billeter MA, Cattaneo R. Molecular biology of defective measles viruses persisting in the human central nervous system. In: Kingsbury DW, ed. The Paramyxoviruses. New York: Plenum Press, 1991: 323-345.

47. Bass BL, Weintraub H, Cattaneo R, Billeter MA. Biased hypermutation of viral RNA genomes could be due to unwinding/modification of double- stranded RNA. Cell 1989; 56:331.

48. Gavish D, Kleinman Y, Morag A, Chajek-Shaul T. Hepatitis and jaundice associated with measles in young adults. Arch Int Med 1983; 143:674-677.

49. Triger DR, Kurtz JB, MacCallum FO, Wright R. Raised antibody titres to measles and rubella viruses in chronic active hepatitis. Lancet 1972; 1:665-667.

50. Randall RE, Russell WC. Paramyxovirus persistence. Consequences for host and virus. In: Kingsbury DW, ed. The Paramyxoviruses. New York: Plenum Press, 1991: 299-321.

51. Robertson DAF, Zhang SL, Guy EC, Wright R. Persistent measles virus genome in autoimmune chronic active hepatitis. Lancet 1987; 2:9-11.

52. Kalland K-H, Endresen C, Haukenes G, Schrumpt E. Measles-specific nucleotide sequences and autoimmune chronic active hepatitis. Lancet 1989; 1:1390-1391.

53. ter Meulen V, Stephenson JR. The possible role of virus infections in multiple sclerosis and other demyelinating diseases. In: Haupike JF, Adams CWM, Rourtellotte WW, eds. Multiple Sclerosis: Pathology, Diagnoses and Management. Baltimore: Williams and Wilkins, 1981:379-399.

54. Adams JM, Imagawa DT. Measles antibodies in multiple sclerosis. Proc Soc Exp Biol Med 1962; 111:562-566.

55. Haire M, Frazer KB, Miller JHD. Measles and other virus specific immunoglobulins in multiple sclerosis. Brit Med J 1973; 3:612-615.

56. Norrby E, Link H, Olsson JE et al. Comparison of antibodies against different viruses in cerebral fluid and serum samples from patients with multiple sclerosis. Inf Imm 1974; 10:688-694.

57. Cook SD, Dowling PC, Russell WC. Multiple sclerosis and canine distemper. Lancet 1978; 1:605-606.

58. Hughes RAC, Russell WC, Froude JRL, Jarrett RJ. Pet ownership, distemper antibodies and multiple sclerosis. J Neurol Sci 1980; 47:429-432.

59. Goswami KKA, Randall RE, Lange LS, Russell WC. Antibodies against the paramyxovirus SV5 in the cerebral spinal fluids of some multiple sclerosis patients. Nature 1987; 327:244-247.

60. Stevens JG, Bastone VB, Ellison GW, Myers LW. No measles virus genetic information detected in multiple sclerosis derived brains. Ann Neurol 1980; 8:625-627.
61. Dowling PC, Blumberg BM, Kolakofsky D et al. Measles virus nucleic acid sequences in human brain. Vir Res 1986; 5:97-107.
62. Haase AT, Ventura P, Gibbs CJ, Tourtellotte WW. Measles virus nucleotide sequences: detection by hybridisation in situ. Science 1981; 212:672-675.
63. Haase AT, Stowring L, Ventura P et al. Detection by hybridisation of viral infection of the CNS. Ann New York Acad Sci 436:103-108.
64. Cosby SL, McQuaid S, Taylor MJ et al. Examination of eight cases of multiple sclerosis and 56 neurological and non-neurological controls for genomic sequences of measles virus, canine distemper virus, simian virus 5 and rubella virus. J Gen Virol 1989; 70: 2027-2036.
65. Wakefield AJ, Sawyerr AM, Dhillon AP et al. Pathogenesis of Crohn's disease: Multifocal gastrointestinal infarction. Lancet 1989; 2:1057-1062.
66. Wakefield AJ, Pittilo RM, Sim R et al. Evidence of persistent measles virus infection in Crohn's disease. J Med Virol 1993; 39:345-353.
67. Rebel A, Malkani K, Basle M. Anomalies nucleaires des osteoclasts de la maladie osseuse de Paget. Nouv Pres Med 1974; 3:1299-1301.
68. Mills BG, Singer FR. Nuclear inclusions in Paget's disease of bone. Science 1976; 194:201-202.
69. Gheradi G, Lo Cascio V, Bonucci E. Fine structure of nuclei and cytoplasm of osteoclasts in Paget's disease of bone. Histopath 1980; 4:63-74.
70. Howatson AF, Fornasier VL. Microfilaments associated with Paget's disease of bone: Comparison with nucleocapsids of measles virus and respiratory syncytial virus. Intervirol 1982; 18:150-159.
71. Mii Y, Miyauchi Y, Honoki K et al. Electron microscopic evidence of a viral nature for osteoclast inclusions in Paget's disease of bone. Virch Arch 1994; 424:99-104.
72. Schajowicz F, Ubios AM, Santini Araujo E, Cabrini RL. Virus-like intranuclear inclusions in giant cell tumour of bone. Clin Orth Rel Res 1985; 201:247-250.
73. Beneton MNC, Harris S, Kanis JA. Paramyxovirus-like inclusions in two cases of pycnodysostosis. Bone 1987; 8:211-217.
74. Mills BG, Yabe H, Singer FR. Osteoclasts in human osteopetrosis contain viral-nucleocapsid-like nuclear inclusions. J Bone Min Res 1988; 3:101-106.
75. Dickson GR, Shirodria PV, Kanis JA et al. Familial expansile

osteolysis. A morphological, histomorphometric and serological study. Bone 1991; 12:331-338.

76. Bianco P, Silvestrini G, Ballanti P, Bonucci E. Paramyxovirus-like nuclear inclusions identical to those of Paget's disease of bone detected in giant cells of primary oxalosis. Virch Arch 1992; 421A:427-433.

77. Rebel A, Basle M, Pouplard A et al. Viral antigens in osteoclasts from Paget's disease of bone. Lancet 1980; 2:344-346.

78. Singer FR, Mills BG. Evidence for a viral aetiology of Paget's disease of bone. ClinOrth Rel Res 1983; 178:245-251.

79. Basle MF, Russell WC, Goswami KKA et al. Paramyxovirus antigens in osteoclasts from Paget's bone tissue detected by mononuclear antibodies. J Gen Virol 1985; 66:2103-2110.

80. Mills BG, Singer FR, Weiner LP, Holst PA. Immunohistological demonstration of respiratory syncytial virus antigens in Paget's disease of bone. Proc Nat Acad Sci USA 1981; 78:1209-1213.

81. Pringle CR, Wilkie ML, Elliott RM. A survey of respiratory syncytial virus and parainfluenza virus type 3 neutralising and immunoprecipitating antibodies in relation to Paget's disease. J Med Virol 1985; 17:377-386.

82. Mills BG, Singer FR, Weiner LP et al. Evidence for both respiratory syncytial virus and measles virus antigens in the osteoclasts of patients with Paget's disease of bone. Clin Orth Rel Res 1984; 183:303-311.

83. Basle MF, Fournier JG, Rozenblatt S et al. Measles virus RNA detected in Paget's disease bone tissue by in situ hybridisation. J Gen Virol 1986; 67:907-913.

84. Gordon MT, Anderson DC, Sharpe PT. Canine distemper virus localised in bone cells of patients with Paget's disease. Bone 1991; 12:195-201.

85. Cartwright EJ, Gordon MT, Freemont AJ et al. Paramyxoviruses and Paget's disease. J Med Virol 1993; 40:133-141.

86. Ralston SH, Digiovine FS, Gallacher SJ et al. Failure to detect paramyxovirus sequences in Paget's disease of bone using the polymerase chain reaction. J Bone Min Res 1991; 6:1243-1248.

87. Birch MA, Taylor W, Fraser WD et al. Absence of paramyxovirus RNA in cultures of pagetic bone cells and in pagetic bone. J Bone Min Res 1994; 9:11-16.

88. Nuovo MA, Nuovo GJ, MacConnell P et al. In situ analysis of Paget's disease of bone for measles-specific PCR amplified cDNA. Diagn Mol Path 1992; 1:256-265.

89. Roodman GD. Biology of the osteoclast in Paget's disease. Sem Arth Rheum 1994; 23:235-236.

90. Reddy SV, Singer FR, Roodman GD. Marrow mononuclear cells

from patients with Paget's disease contain measles virus nucleocapsid mRNA that have mutations in a specific region of the sequence. J Clin Endocrinol Met 1995; 80:2108-2111.

91. O'Driscoll JB, Anderson DC. Past pets and Paget's disease. Lancet 1985; 2:919-921.

92. O'Driscoll JB, Buckler HM, Jeacock J, Anderson DC. Dogs, distemper and osteitis deformans: A further epidemiological study. Bone Min 1990; 11:209-216.

93. Holdaway IM, Ibbertson HK, Wattie D et al. Previous pet ownership and Paget's disease. Bone Min 1990; 8:53-58.

94. Barker DJP, Detheridge FM. Dogs and Paget's disease. Lancet 1985; 2:1245.

95. Siris ES, Kelsey JL, Flaster E, Parker S. Paget's disease of bone and previous pet ownership in the United States: Dogs exonerated. Int J Epidemiol 1990; 19:455-458.

96. Gordon MT, Mee AP, Anderson DC, Sharpe PT. Canine distemper virus transcripts sequenced from pagetic bone. Bone Min 1992; 19:159-174.

97. Mee AP, Sharpe PT. Dogs, distemper and Paget's disease. BioEssays 1993; 15:783-789.

98. Mee AP, Sharpe PT. Letter to the editor. Bone Min 1994; 24:75-76.

99. Morgan-Capner P, Robinson P, Clewley G et al. Measles antibody in Paget's disease. Lancet 1981; 2:733.

100. Winfield J, Sutherland S. Measles antibody in Paget's disease. Lancet 1981; 1:891.

101. Basle MF, Kouyoumdjian S, Pouplard A et al. Paget's bone disease. Preliminary serological study. Pathol Biol 1982; 31:41-44.

102. Hamill RJ, Baughn RS, Mallette LE et al. Serological evidence against the role of canine distemper virus in the pathogenesis of Paget's disease of bone. Lancet 1986; 2:1399.

103. Gordon MT, Bell SC, Mee AP et al. The prevalence of canine distemper antibodies in the pagetic population. J Med Virol 1993; 40:313-317.

104. Jenner E (cited by Kirk H). Canine Distemper, its Complications, Sequelae and Treatment, 1st ed. London: Baillere, Tindall and Cox, 1922.

105. Carré H. Sur la maladie des jeunes chiens. Comp Rend Acad Sci 1905; 140:689-690, 1489-1491.

106. Yoshikawa Y, Ochikubo F, Matsubara Y et al. Natural infection with canine distemper virus in a Japanese Monkey (Macaca fuscata). Vet Microbiol 1989; 20:193-205.

107. Visser IKG, Kumarev V, Orvell C et al. Comparison of two morbilliviruses isolated from seals during outbreaks of distemper in North West Europe and Siberia. Arch Virol 1990; 111:149-164.

108. Nicolle C. La maladie du jeune âge des chiens est transmissible expérimentalement à l'homme sous forme inapparente. Arch Inst Pasteur Tunis 1931; 20:321-323.

109. Bürge T, Griot C, Vandevelde M, Peterhans E. Antiviral antibodies stimulate production of reactive oxygen species in cultured canine brain cells infected with canine distemper virus. J Virol 1989; 63:2790-2797.

110. Vandevelde M, Kristensen B, Braund KG et al. Chronic canine distemper virus encephalitis in mature dogs. Vet Path 1980; 17:17-29.

111. MacIntyre AB, Trevan DJ, Montgomerie R. Observations on canine encephalitis. Vet Rec 1948; 60:635-642.

112. Bodingbauer J. Retention of teeth in dogs as a sequel to distemper infection. Vet Rec 1960; 72:636-637.

113. Cordy DR. Canine encephalomyelitis. Cornell Vet 1942; 32:11-28.

114. Lincoln SD, Gorham JR, Ott RL, Hegreberg GA. Aetiolgic studies of old dog encephalitis. I. Demonstration of canine distemper viral antigen in the brain in two cases. Vet Path 1971; 8:1-8.

115. Lincoln SD, Gorham JR, Davis WC, Ott RL. Studies of old dog encephalitis. II. Electron microscopic and immunohistologic findings. Vet Path 1973; 10:124-129.

116. Imagawa DT, Howard EB, Van Pelt LF et al. Isolation of canine distemper virus from dogs with chronic neurological diseases. Proc Soc Exp Biol Med 1980; 164:355-362.

117. Bennett D. Immune based inflammatory joint disease of the dog. Canine rheumatoid arthritis 1. Clinical, radiological and laboratory investigations. J Small Anim Prac 1987a; 28:779-798.

118. Bennett D. Immune based inflammatory joint disease of the dog. Canine rheumatoid arthritis 2. Pathological investigations. J Small Anim Prac 1987b; 28:799-820.

119. Bell SC, Carter SD, Bennett D. Canine distemper viral antigens and antibodies in dogs with rheumatoid arthritis. Res Vet Sci 1991; 50:64-68.

120. May C, Carter SD, Bell SC, Bennett D. Immune responses to canine distemper virus in joint diseases of dogs. Brit J Rheum 1993; 33:27-31.

121. Schumacher HR, Newton C, Halliwell EW. Synovial pathologic changes in spontaneous canine rheumatoid-like arthritis. Arth Rheum 1980; 23:412-423.

122. Dryll A, Cazalais P, Ryckewaert A. Lymphocyte tubular structures in rheumatoid arthritis. J Clin Path 1977; 30:822-826.

123. Gottlieb NL, Ditchek N, Poiley J, Kiem IM. Pets and rheumatoid arthritis. An epidemiologic survey. Arth Rheum 1974; 17:229-234.

124. Mee AP, Webber DM, May C et al. Detection of canine distem-

per virus in bone cells in the metaphyses of distemper-infected dogs. J Bone Min Res 1992; 7:829-834.

125. Baumgartner W, Boyce RW, Weisbrode SE et al. Histological and immunocytochemical characterisation of canine distemper-associated metaphyseal bone lesions in young dogs following experimental infection. Vet Path 1995; 32:702-709.

126. Meier H, Clark ST, Schnelle GB, Will DH. Hypertrophic osteodystrophy associated with disturbance of vitamin C synthesis in dogs. J Am Vet Med Assoc 1957; 130:483-491.

127. Grøndalen J. Metaphyseal osteopathy (hypertrophic osteodystrophy) in growing dogs. A clinical study. J Small Anim Prac 1976; 17:721-735.

128. Woodard JC. Canine hypertrophic osteodystrophy, a study of the spontaneous disease in littermates. Vet Path 1982; 19:337-354.

129. Merillat LA. Barlow's disease of the dog. Vet Med 1936; 31:304-306.

130. Holmes JR. Suspected skeletal scurvy in the dog. Vet Rec 1962; 74:801-813.

131. Watson ADJ, Blair RC, Farrow BRH et al. Hypertrophic osteodystrophy in the dog. Aust Vet J 1973; 49:433-439.

132. Rendano VT, Dueland R, Sifferman RL. Letter to the editor: Metaphyseal osteopathy: (hypertrophic osteodystrophy). J Small Anim Prac 1977; 18:679-683.

133. Vaananen M, Wikman L. Scurvy as a cause of osteodystrophy. J Small Anim Prac 1979; 20:491-500.

134. Bennett D. Nutrition and bone disease in the dog and cat. Vet Rec 1976; 98:313-320.

135. Teare JA, Krook L, Kallfelz FA, Hintz HF. Ascorbic acid deficiency and hypertrophic osteodystrophy in the dog: a rebuttal. Cornell Vet 1979; 69:384-401.

136. Hedhammar A, Wu FM, Krook L et al. Oversupplementation and skeletal disease: an experimental study in growing Great Dane dogs. Cornell Vet 1974; 64(Suppl. 5):11-160.

137. Hazewinkel HAW, Goedegebuure SA, Poulos PW, Wolvekamp WTC. Influences of chronic calcium excess on the skeletal development of growing Great Danes. J Am Anim Hosp Assoc 1985; 21:377-391.

138. Goedegebuure SA, Hazewinkel HAW. Morphological findings in young dogs chronically fed a diet containing excess calcium. Vet Path 1986; 23:594-605.

139. Mee AP, Gordon MT, May C et al. Canine distemper virus transcripts etected in the bone cells of dogs with metaphyseal osteopathy. Bone 1993; 14: 59-67.

140. Mee AP, May C, Bennett D et al. Generation of multinucleated

osteoclast-like cells from canine bone marrow: Effects of canine distemper virus. Bone 1995; 17:47-55.

141. Kukita A, Chenu C, McManus LM et al. Atypical multinucleated cells form in long-term marrow cultures from patients with Paget's disease. J Clin Invest 1990; 85: 1280-1286.

142. Demulder A, Takahashi S, Singer FR et al. Abnormalities in osteo-clast precursors and marrow accessory cells in Paget's disease. Endocrinol 1993; 1978-1982.

143. Mee AP, Hoyland JA Baird P et al. canine bone marrow cell cultures infected with canine distemper virus: An in vitro model of Paget's disease. Bone 1995; 17:461S-466S.

144. Hoyland JA, Freemont AJ, Sharpe PT. Interleukin-6 (IL-6), IL-6 receptor and IL-6 nuclear factor gene expression in Paget's disease. J Bone Min Res 1994; 9:75-80.

145. Hoyland JA, Sharpe PT. Up-regulation of c-Fos protooncogene expression in pagetic osteoclasts. J Bone Min Res 1994; 9: 1191-1194.

CHAPTER 5

CYTOKINES AND GROWTH FACTORS IN PAGET'S DISEASE

Mark A. Birch and James A. Gallagher

INTRODUCTION

Paget's disease is a focal disorder of bone in which there is a localized increase in bone remodeling. The cellular actions which constitute normal bone remodeling are complex and require the spatially and temporally coordinated actions of several cell types.[1] Remodeling commences with an initiation signal(s) leading to a phase of resorption where bone is excavated from the remodeling site by multinucleated osteoclasts. In pagetic foci, but not in unaffected sites, the osteoclasts exhibit increased multinuclearity and are overactive, resulting in excessive resorption. Normally resorption lacunae are then colonized by osteoblasts which synthesize bone matrix components (osteoid) and the bone is replaced. The mechanism(s) which ensures that the processes of bone resorption and formation remain tightly linked is termed "coupling."[2] The increased osteoclastic activity observed in Paget's disease is matched by an increase in osteoblastic activity, so bone resorption and bone formation remain coupled.

Although systemic hormones such as parathyroid hormone (PTH) clearly have a role in the regulation of bone turnover, the control of remodeling requires paracrine and autocrine interactions between cells at the remodeling locus.[1] The focal nature of Paget's

The Molecular Biology of Paget's Disease, edited by Paul T. Sharpe.
© 1996 R.G. Landes Company.

disease indicates that the pathological mechanisms underlying the disorder lead to perturbations in the local rather than systemic regulation of remodeling. The local regulation of bone remodeling has been the subject of intensive investigation. A wide range of molecules have been shown to influence bone remodeling, including prostaglandins, free radicals and purinergic agonists, but the most extensively studied factors are the cytokines and growth factors. It is clear that cytokines and growth factors can be produced within bone and have significant effects on the actions of bone cells[3,4] and it is now widely accepted that these molecules play a central role in the regulation of remodeling. The pivotal role and complexity of the cytokine network offers a multitude of potential sites where breakdown of normal regulatory mechanisms could lead to the pathological remodeling observed in Paget's disease. In this chapter we will discuss the role that cytokines and growth factors play in normal remodeling and review the evidence implicating them in the pathogenesis of Paget's disease.

CYTOKINES AND GROWTH FACTORS IN NORMAL BONE REMODELING

Cytokines and growth factors can be defined as soluble (glyco)proteins released by cells which act at pico/nanomolar concentrations to nonenzymatically regulate cell function.[5] They normally act locally but in pathological situations, e.g., multiple myeloma, overexpression can result in systemic effects. Many cytokines were originally described as lymphocyte-derived mediators and defined in terms of the system in which they were characterized. Consequently cytokines were thought to be produced by one cell type; to act only as a stimulator or inhibitor; and any other action was related to the principal function in an obvious way. These assumptions about cytokines and growth factors are now known to be flawed. Most can be synthesized by a wide range of unrelated cell types and have diverse bioactivities, including promotion and inhibition of cellular proliferation and differentiation. In addition, cytokines and growth factors often exhibit considerable functional overlap, with one factor being able to replace another in a given situation. This complex network of overlapping actions must be considered when contemplating the role cytokines and growth factors play in bone remodeling.

CYTOKINES AND GROWTH FACTORS WHICH REGULATE BONE RESORPTION

The osteoclast is responsible for the resorption of bone and achieves this by a combination of acidification to demineralize the bone and proteolysis to break down matrix proteins.[6] Control of osteoclastic resorption can occur at two distinct points: firstly by the induction of progenitor cells to form osteoclasts, and secondly, by the activation of mature osteoclasts. Cytokines and growth factors have been shown to have effects on both of these processes (Fig. 5.1).

REGULATION OF OSTEOCLASTOGENESIS

Interleukin-1 (IL-1) was the first cytokine shown to have profound effects on bone remodeling and remains the most potent in vitro with maximal activity at 10^{-10}M. [7] In vitro models of bone resorption, such as mouse calvariae cultures, have shown that IL-1 increases the numbers of osteoclasts after 48 hours. The kinetics of the response to IL-1 revealed that it was required for as little as an hour in the cultures to show a significant effect on osteoclast numbers after 72 hours. There was however no detectable increase in the numbers of osteoclasts before 24 hours. The effects of IL-1 in an in vitro model of osteoclast formation have also been studied.[8] In long-term culture of marrow cells, where multinucleate cells with many of the characteristics of osteoclasts form, IL-1 stimulates the formation of these cells over a similar concentration range to the calvariae studies. This has led investigators to propose that IL-1 mediates its effect on bone resorption by increasing the number of osteoclasts.

The tumor necrosis factors (TNF) α and β have a similar effect to IL-1 on bone resorption. TNFα and β stimulate resorption in organ culture[9] and enhance formation of multinucleate cells in marrow cultures;[8] however. their potency is much lower than IL-1, being in the range 10^{-9} to 10^{-7} M. TNFα seems in most assays of bone resorption to be more potent than TNFβ.

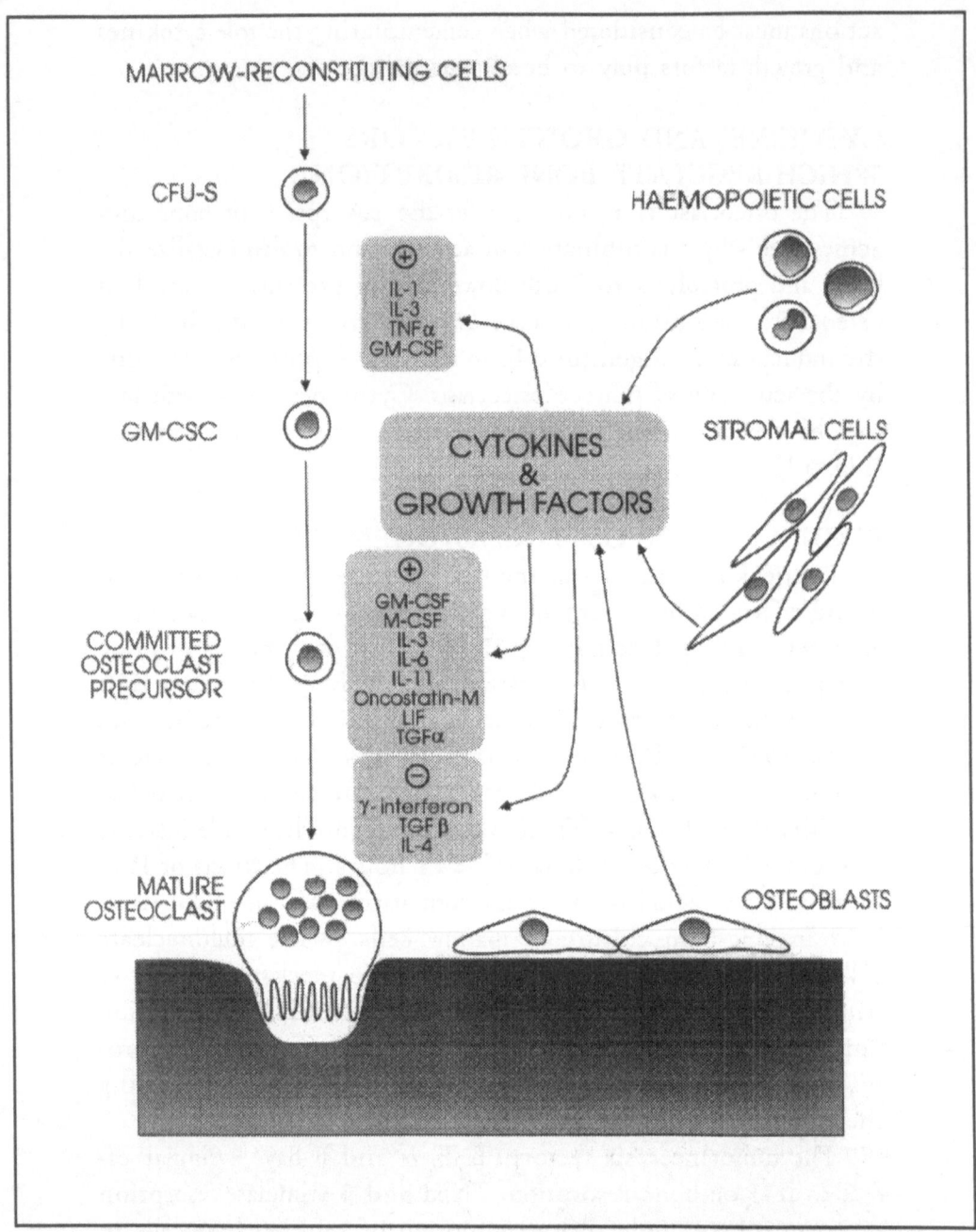

Fig. 5.1. *The role of cytokines and growth factors in the formation and function of the osteoclast. Cytokines and growth factors derived from a number of cellular sources, including hemopoietic cells; stromal cells and osteoblasts have profound effects on osteoclastogenesis and mature osteoclast activity. Some may act by stimulating the maturation of CFU-S through to GM-CSF, others on the formation of osteoclasts from the committed precursor whilst some factors affect all stages of osteoclastogenesis. Abnormal production of, or increased sensitivity to, cytokines and growth factors may be responsible for the increased numbers, hypernuclearity and increased activity of osteoclasts in Paget's disease.*

The osteoclast is derived from a hemopoietic precursor through a similar lineage to monocytes and macrophages. Maturation proceeds from undifferentiated stem cells, which can be both self renewing (totipotent) and nonself renewing (pluripotent), through cells committed to the osteoclast lineage to the mature osteoclast itself. These changes are controlled by colony stimulating factors (CSFs) along with other factors, such as $1,25(OH)_2D_3$. There are three CSFs, IL-3, granulocyte-macrophage colony stimulating factor (GM-CSF) and macrophage colony stimulating factor (M-CSF), which have been shown to affect osteoclastogenesis (Fig. 5.1).

IL-3 induces the differentiation and proliferation of early pluripotent stem cells and committed progenitors. When IL-3 treated murine stem cells are added to fetal long-bone cultures there is increased hemopoiesis and osteoclast number.[10] When murine bone marrow cells were cultured with wafers of devitalized bovine bone, IL-3 did not induce bone resorption.[11] Addition of $1,25(OH)_2D_3$ to these cultures did cause the formation of bone-resorbing cells and in greater numbers than cultures treated with $1,25(OH)_2D_3$ alone. A similar study failed to observe this effect and in fact IL-3 reduced the number of multinucleate cells induced by $1,25(OH)_2D_3$.[12] In general though, IL-3 is thought to induce the proliferation and differentiation of stem cells and more mature osteoclast progenitors. It does not, however, support the full maturation of the osteoclast, with other factors such as $1,25(OH)_2D_3$ required to produce bone-resorbing cells.

GM-CSF acts on cells which have differentiated from stem cells, but are still capable of multilineage development, and induces their proliferation and maturation into granulocytes, macrophages, megakaryocytes and erythrocytes. The role of GM-CSF in osteoclast development is complex. GM-CSF alone and with $1,25(OH)_2D_3$ has been shown to increase the numbers of multinucleate cells formed in marrow cultures.[13] Other studies, however, have shown that addition of GM-CSF to long-bone cultures reduces the rate of osteoclast formation.[14] Moreover, incubation of GM-CSF-treated marrow cells with slices of devitalized bone resulted in the reduction of bone resorption.[11] One interpretation of this data is that addition of GM-CSF to a more differentiated progenitor drives the cell away from the osteoclast lineage towards macrophage and granulocyte development. What is clear, however, is that GM-CSF does not directly enhance bone resorption but does have effects on osteoclast formation.

Of all the CSFs, M-CSF has the most profound effects on osteoclastogenesis. In long-term marrow cultures M-CSF induces the formation of multinucleate cells.[13] When added to mouse fetal long-bone cultures,[15] M-CSF stimulates bone resorption, however this effect is not repeated when mouse marrow cells are cultured on wafers of devitalized bone.[11] Perhaps the clearest evidence of the role M-CSF plays in bone resorption was the discovery that bone defect observed in the *op/op* mouse can be traced to a mutation in the *M-CSF* gene.[16] As with GM-CSF there is conflicting data, but it seems that M-CSF, probably acting with other factors, supports the later stages of osteoclast development.

IL-6 is an important mediator of immune and hemopoietic cells but also plays a role in osteoclastogenesis. Most of the evidence suggests that IL-6 stimulates osteoclast formation from precursors in a manner analogous to GM-CSF. IL-6 increases the formation of multinucleate cells in both untreated and $1,25(OH)_2D_3$-treated human long-term marrow cultures.[17] In mouse metacarpal cultures, where there are few mature osteoclasts, IL-6 significantly enhanced bone resorption.[18] This was not repeated in cultures of 17 day old fetal mouse radii, which contain higher numbers of mature osteoclasts. Some studies have demonstrated that IL-6 increases osteoclast formation in calvariae cultures,[19] while in others an inhibition has been shown.[20,21] In vivo evidence for IL-6 increasing the number and activity of osteoclasts was shown in experiments where nude mice bearing cells transfected with the IL-6 gene developed a number of conditions, including hypercalcemia.[22] These studies show that IL-6 predominantly exerts its effect on bone resorption by stimulating osteoclast maturation.

Leukemia inhibitory factor (LIF) was initially identified by its ability to suppress the proliferation and induce the differentiation of myeloid leukemia cells. In mouse calvaria cultures LIF increased the number and activity of osteoclasts.[23] In addition, when LIF was added to marrow cultures the formation of multinucleate cells was increased dose-dependently.[24] These results again suggest that LIF is important in the differentiation of the osteoclast.

IL-11 was discovered as a cross-reacting peptide in an IL-6 bioassay and shares many of the biological properties of that cytokine. Recent studies have demonstrated that it stimulates osteoclast formation and that it may well mediate the osteoclastogenic effects of PTH, IL-1β and TNFα.[25]

Several studies using the long-term marrow culture system have revealed that γ-interferon,[26] IL-4[27] and TGFβ[28] inhibit the formation of multinucleate cells. In addition, γ–interferon has been shown to preferentially inhibit cytokine-stimulated resorption in organ cultures.[29] IL-4 has been shown to inhibit PTH, $1,25(OH)_2D_3$ and IL-1 stimulated bone resorption in organ culture.[30] It has been proposed that γ–interferon, IL-4 and TGFβ prevent both the proliferation and differentiation of osteoclast precursors.

Osteoclastogenesis is a complex process which is tightly controlled by a number of factors including cytokines and growth factors. Some molecules, such as IL-1 and IL-3, influence the primitive stem and progenitor cells, stimulating proliferation and development and often preparing them for later exposure to other stimuli. Other factors, including GM-CSF and M-CSF, act at specific points in the lineage. All of the factors, however, have significant effects on the rate of bone resorption.

ACTIVATION OF MATURE OSTEOCLASTS

The activity of preformed or mature osteoclasts can be assessed in vitro by using a bone slice assay. Rodent or chick marrow cells are cultured on wafers of devitalized bovine bone and authentic osteoclast excavations (resorption lacunae) in the bone surface can be quantitated by microscopy. In these assays both IL-1[31] and TNF[32] stimulate bone resorption and each augments the action of the other. Also IL-1 and TNF when added together enhance the effect of the other[33] and interact synergistically with PTH and PTH-related peptide (PTHrP).[34] The stimulation of osteoclastic bone resorption by PTH and PTHrP is recognized to be mediated through the PTH-receptor positive osteoblast. In response to PTH-receptor activation, the bone resorption is induced either by osteoblast production of an as yet unidentified factor which acts directly on the osteoclast or by the osteoblast peeling away from the bone allowing the osteoclast access to the bone surface.[35] IL-1 and

TNF are also thought to stimulate bone resorption in this manner, but no convincing intermediary between the osteoblast and the osteoclast has been identified.

The activity of mature osteoclasts can be directly inhibited by some cytokines and growth factors. TGFβ is directly released from the bone matrix and activated from its latent form into the mature peptide by the acidic environment of the resorption site.[36] It is thought to decrease superoxide production and lower levels of tartrate-resistant acid phosphatase and reduce the activity of the osteoclast. TGFβ has profound effects on the osteoblasts (see later), so down-regulation of mature osteoclast activity may also be mediated by an as yet undiscovered mechanism through these cells. In addition, this may also be a common mechanism of action for other osteoclast inhibiting cytokines, such as IL-4.

It has been reported that a pentapeptide fragment of PTHrP, corresponding to amino acids 107-111, also directly inhibits osteoclast activity.[37,38] Other workers, however, have been unable to demonstrate this effect of PTHrP[107-111] in mouse calvarial cultures.[39] So, in addition to stimulating resorption through its PTH-like actions on osteoblasts, the PTHrP molecule may also have the potential to inhibit osteoclasts.

Bone resorption can be regulated by cytokines and the most potent stimulators of bone resorption, IL-1 and TNF, affect not only osteoclastogenesis but also mature osteoclast activity as well. In addition, TGFβ and possibly PTHrP[107-111] may directly inhibit bone resorption by acting on the osteoclast.

BONE FORMATION

Following a phase of resorption in normal bone remodeling, new bone is laid down, fully replacing the matrix which had previously been removed. This requires the recruitment and differentiation of osteoblast precursors followed by bone matrix production and mineralization by the mature osteoblast. Cytokines and growth factors have profound effects on both of these processes (Fig. 5.2).

OSTEOBLAST PROLIFERATION/DIFFERENTIATION

A population of cells with many of the characteristics of osteoblasts will grow from human trabecular bone which is seeded

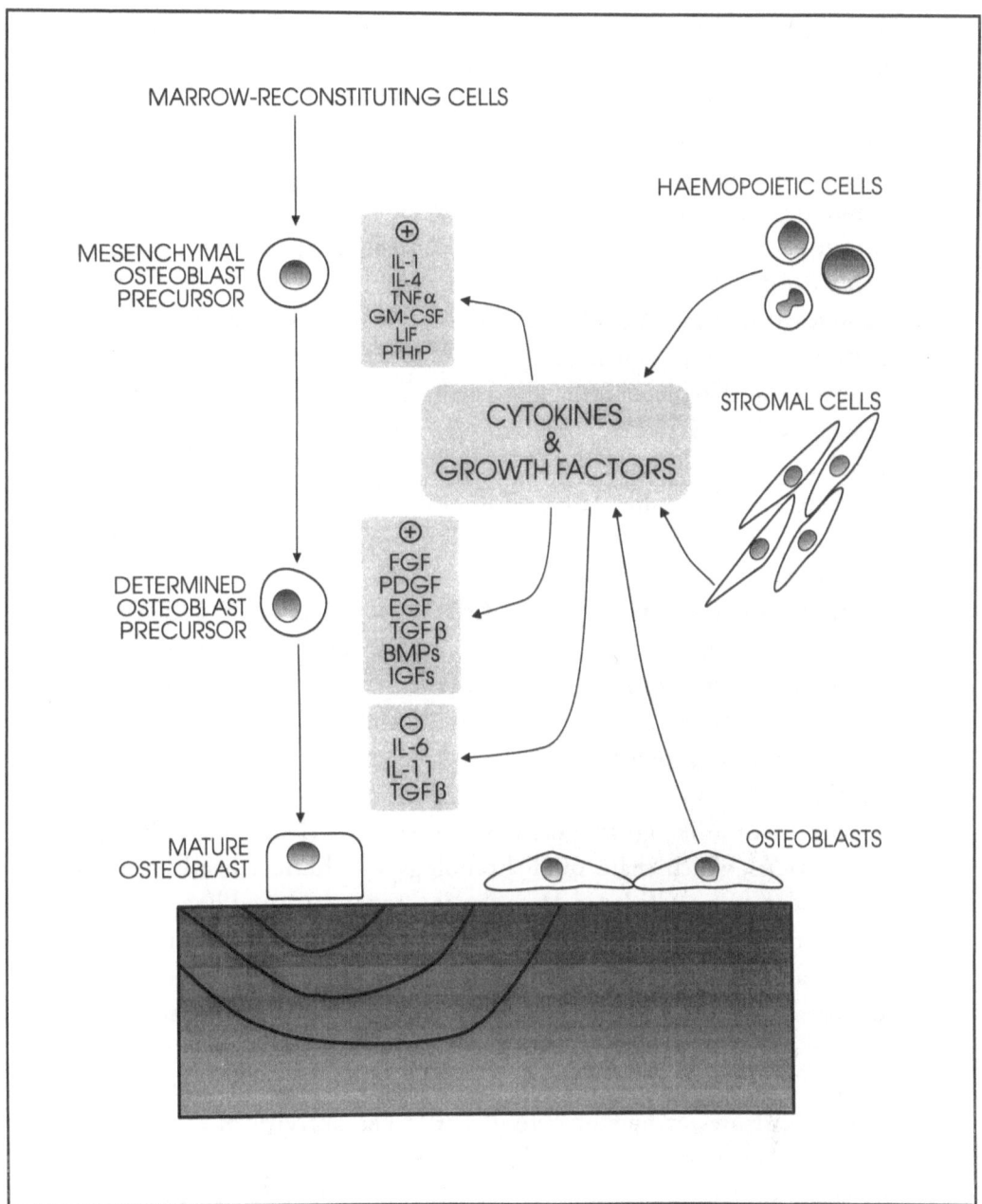

Fig. 5.2. The action of cytokines and growth factors on the maturation of osteoblasts. Cytokines and growth factors, which can be produced locally within the bone microenvironment, play significant roles in the maturation of the osteoblast. Many of the factors which influence osteoclast formation/function are also involved in the regulation of osteoblasts, and are good candidates for involvement in coupling bone resorption to bone formation. The preservation of coupling in Paget's disease may reflect the over/under expression of a single factor regulating resorption and formation. Alternatively the increased resorptive activity in Paget's disease may lead to a subsequent elevation of cytokines and growth factors which influence osteoblastic activity.

into culture. At low concentrations IL-1[40,41] and TNF[42] have been shown to stimulate the proliferation of osteoblasts in cultures of human trabecular bone cells. Although these cultures are heterogenous, IL-1 has been shown to induce the proliferation of cells specifically identified as osteoblasts.[43] In these experiments [^{3}H]-thymidine and IL-1 were added and osteoblast differentiation subsequently induced with $1,25(OH)_2D_3$. Osteoblastic cells which had been stimulated to proliferate by IL-1 were identified by the dual localization of [^{3}H]-thymidine incorporation and expression of the enzyme, alkaline phosphatase.

Acidic and basic fibroblast growth factors (FGFs), which are angiogenic factors considered important in wound healing, also stimulate bone cell replication. In organ culture, addition of FGFs has been shown to stimulate production of collagen, not through effects on the mature osteoblast, but by increasing the number of osteoprogenitors.[44]

Platelet-derived growth factor (PDGF) is a two chain polypeptide which is coded for by two separate genes, *PDGF A* and *PDGF B*. Mature PDGF can therefore exist in three forms, as the homodimers AA or BB, or the heterodimer AB. PDGF stimulates bone DNA and protein synthesis, again through bone cell replication.[45]

The insulin-like growth factors (IGFs) I and II are single chain polypeptides which share 60% homology and hence, overlapping biological effects. IGF-I and II are mitogens for cultured bone cells. In fetal rat calvaria, IGF-I is four to seven times more potent than IGF II at increasing [^{3}H]-thymidine incorporation.[46,47] Studies on osteosarcoma cells have demonstrated that the IGFs rapidly stimulate the *c-FOS* gene, a growth related transcription factor.[48] IGF-I and IGF-II both stimulate cell proliferation dose-dependently in primary cultures of human osteoblasts.[49] The activity of the IGFs is tightly regulated by a group of proteins called the insulin-like growth factor binding proteins (IGFBPs). These IGFBPs can act either to enhance or inhibit the activities of the IGFs and these effects must be considered when interpreting studies.

Transforming growth factor (TGF) βs and the bone morphogenetic proteins (BMPs) are members of a "superfamily" of peptides, which share several conserved cysteine residues and form disulphide-linked homodimers of approximately 25 kD. In mammals, three isoforms of TGFβ have been identified, TGFβ1, β2 and β3. TGFβ has been shown to stimulate new bone formation in vivo.[50-52] In cultures of both fetal rat calvariae[53] and isolated calvariae cells,[54] TGFβ stimulates proliferation and inhibits alkaline phosphatase expression. In studies on osteosarcoma cell lines, TGFβ has been shown to sometimes stimulate proliferation (MG-63)[55] and in others, inhibit (ROS 17/2.8 and MC3T3-E1 cells).[56] The BMPs are all capable of inducing bone and cartilage formation using in vivo assay systems.[57-59] The actions of BMPs are complex and probably mediated by many different molecules through several different cell types, but include the proliferation and differentiation of mesenchymal cells to osteoblasts.

Low doses (2-10 ng/ml) of GM-CSF also stimulate the proliferation of cultured osteosarcoma cells[60] and primary human osteoblasts.[61] LIF is also a dose-dependent mitogen for isolated rodent[62] and human bone cells.[63] In addition, mice grafted with tumor cells overexpressing LIF show a significant increase in the number of osteoblasts and elevated bone formation.[64] PTHrP can also stimulate bone formation through its PTH-like actions[65,66] and this in part includes the stimulation of osteoblast proliferation.

The proliferation and differentiation of mesenchymal cells to the osteoblast lineage is regulated by several cytokines (Fig. 5.2). Interestingly, potent stimulators of bone resorption also support this process and may represent a mechanism to provide sufficient numbers of osteoblasts to replace the bone excavations. This is augmented by the action of factors like TGFβ, which inhibits active osteoclasts and new osteoclast formation, and increases bone cell proliferation.

MATURE OSTEOBLAST ACTIVITY

IL-1 and TNF both reduce the production of alkaline phosphatase and osteocalcin (markers of a mature osteoblast phenotype) by nontransformed cells,[41,67,68] whereas experiments with MC3T3 and MG-63 cells showed a stimulation in alkaline phosphatase.[69,70] In long-term primary cultures of rat osteoblasts the formation of bone nodules is reduced when treated with IL-1.[71] The consensus appears to be that while stimulating the number of osteoblast precursors, IL-1 and TNF inhibit their differentiation.

IL-4 is thought to stimulate the maturation and the function of the mature osteoblast. IL-4 inhibits the proliferation of the MG-63 cell line and induces the expression of alkaline phosphatase.[72] Primary cultures of human osteoblasts respond to IL-4 by increasing the rate at which they form nodules of bone, with concomitant elevation in alkaline phosphatase, hydroxyproline and osteoclacin.[73] In similar experiments IL-11 has been shown to inhibit formation of bone nodules in vitro.[74]

The IGFs have significant effects on the function of the mature osteoblast. In rat calvariae, an increase in type I collagen and matrix component synthesis is observed following treatment with IGF-I, which is independent of cell replication.[47,75] In the human osteosarcoma cell line OHS-4, IGF-I and IGF-II induce osteocalcin synthesis when added to the cells in the presence of $1,25(OH)_2D_3$[76] and this observation is repeated in cultures of human bone cells.[77]

It has been shown that the continued presence of TGFβ in long-term mineralizing cultures of fetal rat calvarial cells inhibits both bone formation and markers of mature osteoblast function.[78] Unlike $1,25(OH)_2D_3$, the TGFβ effect is reversible, provided the cells are at an early enough stage of differentiation. It has therefore been proposed that TGFβ stimulates bone formation through the proliferation of committed osteoprogenitor cells, while continued high concentrations of TGFβ inhibit osteoblast differentiation.

Potent stimulators of resorption tend to inhibit the expression of the mature osteoblast phenotype, whereas factors such as IGFs and BMPs stimulate osteoblastic activity. The hypothesis is that this allows a clear demarcation in the phases of resorption and formation. This would also fit with the recent evidence from studies on TGFβ, where high local concentrations would indicate resorption and lower levels would occur only after the termination of osteoclast activity.

PRODUCTION OF CYTOKINES
AND GROWTH FACTORS IN BONE

Cytokines and growth factors can clearly affect the cells which are responsible for the remodeling of bone. To qualify as important, physiologically relevant modulators of bone resorption and formation, cytokines and growth factors must be produced locally.

Furthermore production must be regulated either hormonally or in response to mechanical strain or fracture, in a way that fits with our understanding of how bone remodeling is controlled.

Osteoclasts have been described as originating from hemopoietic precursors and, within the bone marrow, there are numerous cells of hemopoietic or mesenchymal origin which are capable of producing cytokines and growth factors. In the same environment the early proliferation of the mesenchymal cells which go on to form osteoprogenitor cells, also takes place. Clearly these cells have the opportunity to respond to locally produced cytokines and growth factors.

The osteoblast can synthesize a multitude of cytokines and growth factors[4,79-81] (Table 5.1), which clearly have the potential to be autocrine/paracrine regulators. In addition, cytokines and growth factors produced by osteoblasts may be sequestered into bone during a phase of formation. When a subsequent phase of resorption occurs, factors liberated from the bone could directly affect the mature osteoclast. This has been postulated to provide termination signals to the functional osteoclast, e.g., TGFβ and PTHrP(107-111). In addition, cytokines released from the matrix may be important in "coupling" bone resorption to the subsequent step of bone formation.

Within the bone microenvironment there are a number of cells which are capable of producing a range of cytokines and growth factors. Therefore, to ensure successful bone remodeling, the production of these factors must be carefully regulated. Both IL-1β and TNFα have been shown to be produced by osteoblasts and each augments the bone resorbing activities of the other.[33] IL-1β production is stimulated by TNF and endotoxin but not by PTH or by 17β-oestradiol.[82] Similarly, TNFα release by osteoblasts is elevated by IL-1, GM-CSF or LPS, but PTH and $1,25(OH)_2D_3$ have no effect.[83] Production of IL-6 by human osteoblasts can also be upregulated by IL-1, TNFα and LPS, with PTH and $1,25(OH)_2D_3$ again having no effect.[84] Recently, IL-11 production by SaOS-2 human osteosarcoma cells has been shown to be upregulated by IL-1α, TGFβ1, PTH and PTH-related peptide, with no effect being observed following IL-4, γ-interferon or endotoxin treatment.[81] In the same study, osteoblasts cultured from fragments

Table 5.1. *Cytokine and growth factors expressed by human osteoblasts*

Cytokine or Growth Factor	Expression In Vitro	Expression In Situ
IL-1α	–	ND
IL-1β	+	+ (121)
IL-2	–	ND
IL-3	+	ND
IL-4	–	ND
IL-5	–	ND
IL-6	+	+ (120,121)
IL-7	–	ND
IL-8	+	ND
IL-10	–	ND
IL-11	+ (81)	ND
TNFα	+	+[a]
TNFβ	+	ND
M-CSF	–	ND
GM-CSF	–	ND
G-CSF	–	ND
LIF	+[a]	ND
Oncostatin-M	+[a]	ND
SCF	+[a]	ND
EGF	+[a]	ND
FGFβ	+[a]	ND
IGF-I	+[a]	+ (129)
IGF-II	+[a]	+ (129)
PDGFα	+[a]	ND
PDGFβ	–	ND
PTHrP	+ (127)	+[a]
TGFβ1	+	+ (121)
TGFβ2	+	ND
TGFβ3	+	ND
MBP 1A	+[a]	ND
MBP 2	+[a]	chapter 6
MBP 3	+[a]	chapter 6
MBP 4	+[a]	chapter 6

Production of cytokines has been investigated in primary cultures of human osteoblasts and subsequently confirmed in sections of bone by in-situ hybridization. ND - not determined. Where a reference has not been cited the data are described in ref. 77.
[a] - unpublished observation.

of human trabecular bone were only shown to produce IL-11 after addition of TGFβ, no effect was observed with PTH. These examples show the intricate regulation of cytokine production by osteoblasts and their complex interactions with systemic hormones.

Factors associated with the termination of bone resorption and the stimulation of formation seem more likely to be regulated by calciotropic hormones and more independently of other cytokines. Production of TGFβ by cultured human osteoblasts is stimulated by treatment with 17β-estradiol and PTH.[85] This is supported by in vivo evidence from vitamin D deficient animals where the levels of TGFβ in the bone matrix are lower than normal. IGF production is hormonally regulated, with PTH,[86] $1,25(OH)_2D_3$,[77] 17β-oestradiol[87] and growth hormone,[77] all stimulating IGF production by osteoblasts. In addition, TGFβ has been shown to stimulate IGF-I and IGF-II production by osteoblasts, while BFGF suppresses their synthesis.[88]

Cytokine and growth factor action can be further controlled by the regulation of the membrane-bound receptors which recognize these factors and the subsequent intracellular signaling cascade that receptor activation elicits. Functional redundancy amongst cytokines and growth factors is commonplace and this is now better understood through the study of the molecular biology of the receptor systems.[89] Most cytokine receptors appear to be multichain complexes consisting of a private ligand-binding chain and a public class-specific signal transducer. An example of this is the IL-6, IL-11, LIF and oncostatin M receptor, which uses the same signal transducer, gp130. Thus, these cytokines exert similar functions in various tissues, including bone. A common signal transducer has also been discovered in the IL-3/GM-CSF and IL-2/IL-4 systems. Not all cytokines are recognized by receptors which belong to these families. The receptors for IL-1 and TNF do not follow this common structure. The extracellular domain of the IL-1 receptor is composed entirely of immunoglobulin-like domains. For TNF there are two distinct receptors, each sharing little homology with each other, and completely different intracellular signaling domains.

Cytokine receptor binding induces the activation of intracellular tyrosine kinases, e.g., JAK2 in the case of IL-3, IL-6 and GM-CSF. This in turn is followed by the ras-MAP kinase cascade and the activation of a number of transcription factors. Another signal transduction mechanism which may occur exclusively or in parallel is the tyrosine kinase activation of transcription factor(s) in the cytoplasm and their subsequent localization to the nucleus. What is not clear, however, is how cytokines which exert their effects through common signaling pathways can have unique effects. This is probably achieved through a combination of strategies including the variable expression of private receptors and restriction of signaling molecule activation, which could occur either by preferential activation of a subset, or by signal molecule regulation or modification by other stimuli. In this context the situation that a cell finds itself in is crucial to the way in which it will respond to cytokines and growth factors. It is therefore becoming clear that the extracellular matrix (ECM) has a significant role in the way cells behave.[90]

There are several ways in which the ECM can interact with cytokines and growth factors. Adherence to matrix induces many cell types to express cytokines and this response can also be observed with intercellular adhesion. Treatment of cells with cytokines, as previously described, can cause a change in the expression of matrix proteins and also metalloproteinases. Cytokines can also change the pattern of adhesion receptors which a cell displays. It is well established that some cytokines function when bound to the ECM. FGF,[91] GM-CSF[92] and IL-3[93] can all activate their receptors whilst interacting with ECM proteoglycans, whilst TGFβ is bound and inactivated by the core protein of decorin.[94] Several adhesion receptors can also act like cytokine receptors, with for instance, the addition of fibronection to some cells mimicking FGF, PDGF or IL-1.[95] Lastly, and perhaps most importantly, the ECM controls how cells can respond to an array of cytokines.

The ECM reflects local metabolic history and is critical in understanding how past experience can determine the way a cell responds to cytokines. This is particularly emphasized in bone remodeling, where liberation of cytokines from the bone matrix may be an important part of "coupling" bone resorption to formation.

Thus cytokines and growth factors coupled with the ECM can be considered programming for future cell activity. This may be important in Paget's disease since abnormal bone is formed and some studies have identified that pagetic osteoblasts in vitro show a different pattern of protein secretion, including an altered C-telopeptide of collagen type I.[96]

THE ROLE OF CYTOKINES AND GROWTH FACTORS IN ABNORMAL BONE REMODELING

Cytokines and growth factors are locally produced factors which normally act within the immediate environment of the cells of origin. In pathological situations inappropriate production of cytokines can lead to other organs being exposed to their effects, and this is highlighted in diseases of bone.

Patients with myeloma almost always have extensive bone destruction. The malignant plasma cells of myeloma lodged in the marrow cavity secrete cytokines which activate not only local osteoclasts, but osteoclast progenitors and mature osteoclasts throughout the skeleton. The cytokines responsible for this localized bone destruction and more generalized osteoporosis are known to include TNF,[97] IL-1[98] and IL-6.[99] Many solid tumors also metastasize to bone and usually produce destructive osteolytic lesions. There is a common pattern of metastasis with tumors such as breast, prostate and lung frequently metastasizing to bone.[100] Cytokines have occasionally been associated with these tumors, as in some prostate cancers, for instance. High metastatic activity in prostate cancer cells has been correlated with elevated production of FGF, which is a mitogen for osteoblasts and may account for the pathogenic remodelling.[101]

Humoral hypercalcemia of malignancy (HHM) is a clinical syndrome where cancer cells secrete a circulating calcemic factor.[102] This factor has been identified as PTHrP and the condition is thought to be primarily initiated by the PTH-like actions of the molecule on bone resorption and kidney mineral homeostasis.[103] PTHrP expression by a tumor is also associated with a higher metastatic potential.[104,105] It should also be remembered that PTHrP can be considered a "propeptide" as several potential proteolytic fragments of this molecule have been shown to regulate cell activity differently.[103]

In the joints of patients suffering from rheumatoid arthritis, high levels of several cytokines, including IL-1,[106] IL-6,[107] TNFα[108] and M-CSF,[109] can be found. The presence of these factors has been proposed to be the cause of juxta-articular osteoporosis and bony lesions which are classic radiographic markers of the disease.

The focus of many workers' attention has been on trying to understand the role cytokines and growth factors may play in postmenopausal osteoporosis. Following the menopause levels of estrogen fall and the phases of bone resorption and bone formation become "uncoupled." The net result of this is a fall in the bone mineral density and eventual breakdown of skeletal function. Cytokines were first implicated in postmenopausal osteoporosis, when it was demonstrated that peripheral blood monocytes isolated from patients suffering from the disease produced elevated levels of IL-1.[110] This observation has also been repeated for TNFα.[111] In addition monocytes from ovariectomized premenopausal women with low estrogen levels, produce more IL-1 and TNFα, and when stimulated, produce GM-CSF at higher levels than age-matched controls.[112] More recently a direct link to bone cells has been established with some workers able to show that IL-1 and TNFα stimulated IL-6 production by osteoblasts and stromal cells in vitro is inhibited by oestrogen.[113] In vivo studies have demonstrated that in ovariectomized mice there is increased osteoclastogenesis, and that this is mediated by IL-6.[114] However, this may well not be the only mechanism of bone loss following the menopause.

The IGFs are probably critically involved in bone remodeling and several investigators have focused on the changes in IGFs associated with aging. Some studies have shown that there is a lowering of serum IGF-I and IGF binding proteins with age,[115] however, others have failed to demonstrate significant correlations between bone density (or osteoporotic fracture) in elderly men or women and IGF levels.[116] In acquired growth hormone deficiency serum concentrations of IGF-I and IGFBP-3 predict bone density; however, in general IGF-I is not a useful marker of bone formation.[117] The mechanisms associated with postmenopausal and age-related bone loss are probably numerous and quite subtle. Greater understanding of cytokines and growth factors may reveal how these molecules may contribute to postmenopausal osteoporosis and age-associated bone-loss.

CYTOKINES IN PAGET'S DISEASE

One of the earliest pieces of evidence that abnormal or inappropriate production of cytokines might play a role in Paget's disease came from a study by Pioli and colleagues.[118] Peripheral and medullary monocytes were obtained from patients with and without Paget's disease and seeded into culture. After 24 hours the conditioned medium was removed from the cultures and bioassayed for IL-1. In untreated cultures of peripheral cells there was no detectable IL-1, whereas when treated with endotoxin, IL-1 was easily detectable with the levels expressed by the cells from pagetic patients being significantly higher than normals. Medullary monocytes were shown to spontaneously produce IL-1 and again, the level of production was greater when stimulated by endotoxin, with pagetic cells producing significantly higher levels than normal controls.

The production of IL-6 in Paget's disease has been studied by Roodman and co-workers[119] who investigated long-term marrow cultures set up from cells isolated from patients with Paget's disease. The conditioned media from these cultures was shown to induce the size and number of multinucleate cells, which resemble osteoclasts, in long-term cultures of normal human marrow. The conditioned media from the cultures of pagetic cells was assayed for the presence of several candidate cytokines which could be inducing osteoclastogenesis. No IL-1, TNFα or GM-CSF activity was identified and only addition of high concentrations of anti-IL-1 significantly decreased the rate of multinucleate cell formation. In contrast, elevated levels of IL-6 were detected in the conditioned media from the long-term culture of pagetic marrow cells. In addition, anti-IL-6 completely neutralized the effect of pagetic cell conditioned media on normal marrow cultures. Investigation of the serum and marrow revealed twenty times higher levels of IL-6 in Paget's patients compared to normals. There was no correlation between serum levels of IL-6 and alkaline phosphatase in the patients suffering from Paget's disease in this study.

Other workers have localized IL-6, IL-6 receptor (IL-6R) and IL-6 nuclear factor (NF-IL-6) transcripts in bone.[120,121] Osteoblasts in normal remodeling bone and in pagetic bone expressed IL-6, IL-6R and NF-IL-6, with apparent higher levels in pagetic bone.[120] Transcripts for IL-6R and NF-IL-6 could be detected in osteoclasts from normal bone, however, mRNA for IL-6, in addition to IL-6R and NF-IL-6, was found in pagetic osteoclasts.[120] The high level of IL-6 expression in pagetic osteoclasts may explain why they are outside normal regulatory control. The authors propose that viral infection of the pagetic osteoclast activates transcription factors, such as NF-κB and NF-IL-6, which upregulate the transcription of the *IL-6* and *IL-6R* genes leading to increased osteoclastogenesis (Fig. 5.3, and see also chapter 6).

There are several lines of evidence which support the contention that cytokine production can be regulated by virus infection. Patients infected with human T-cell lymphotrophic virus-type I (HTLV-1) often have hypercalcemia and marked bone destruction.[122] IL-1 has been implicated as the causative factor because of its known role in bone remodeling and its production by some infected T lymphocytes.[123] However, another study has identified PTHrP in all HTLV-1 infected patients studied and also showed that the HTLV-1 *trans* activator, the *tax* gene product, could also stimulate *PTHrP* gene transcription.[124] The tax protein can also transcriptionally activate a wide variety of genes including *IL-2*, *IL-2 receptor*, *c-Fos*, *GM-CSF* and *NF-κB*. This is emphasized by experiments in which mice transgenic for the *HTLV-tax* gene[125] display multiple abnormalities including increased numbers of osteoclasts and incorrectly remodeled bone. A recent study described increased IL-11 production by a lung stromal cell line following infection with RSV.[126] In addition, the same group reported that infection with RSV or CDV stimulates IL-11 synthesis by the osteosarcoma cell line, SaOS-2.[81] This is especially interesting since IL-11 and IL-6 share many functional characteristics and the same gp130 signaling subunit.

The focus of research into Paget's disease has been on the abnormally large and overactive osteoclast. Osteoblasts are known to synthesize several cytokines important in bone remodeling for this reason we chose to phenotype the profile of factors produced by

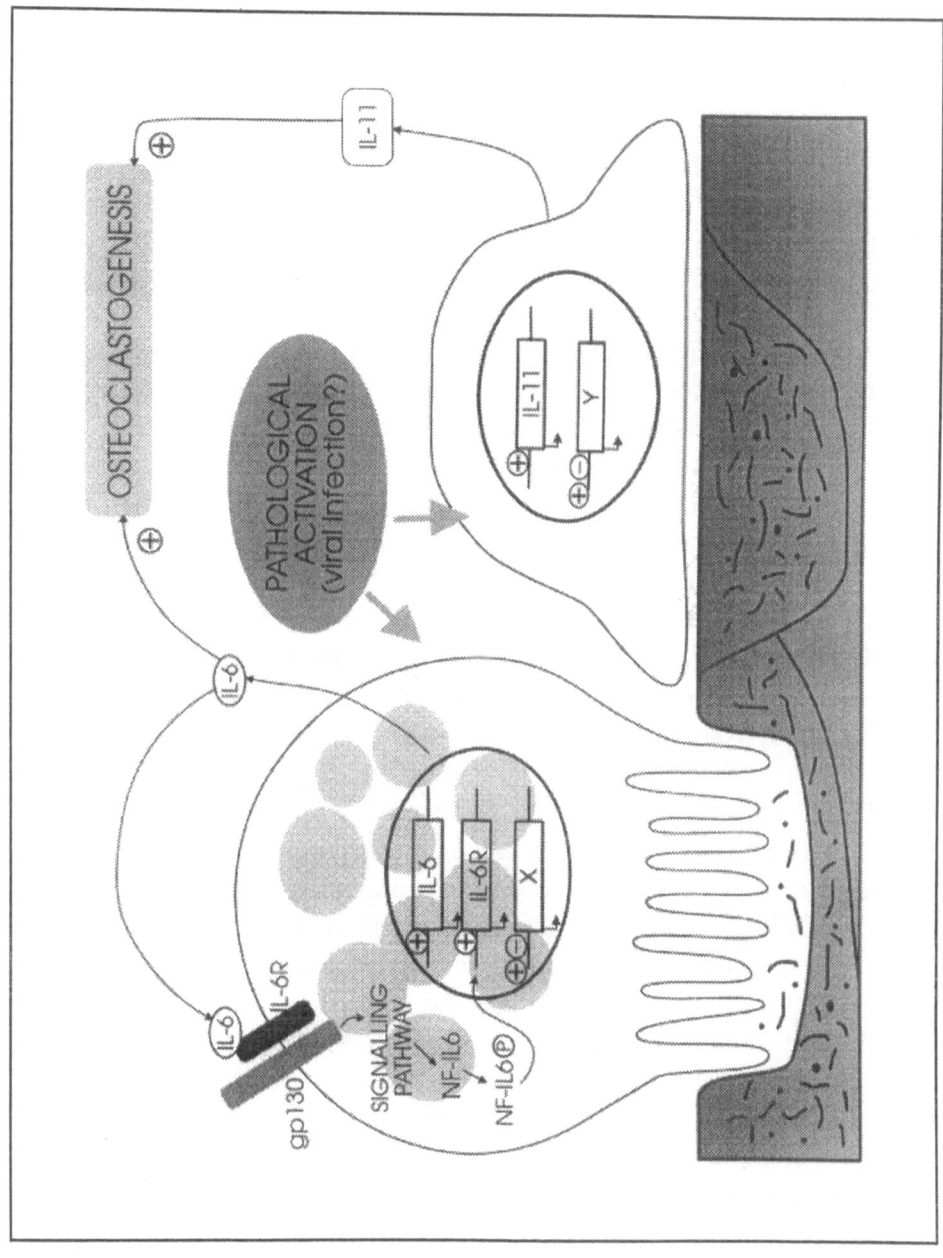

Fig. 5.3 In Paget's disease, IL-6 is reported to be elevated in serum, in marrow biopsies and in long-term marrow cultures. In another study, osteoclasts from patients suffering from Paget's disease expressed transcripts for IL-6, IL-6R and NF-IL-6 whereas in normal osteoclasts only mRNA for IL-6R and NF-IL-6 could be detected. This has led workers to propose that in Paget's disease, production of IL-6 by the osteoclast is increased in response to pathological activation. The elevated levels of IL-6 stimulate osteoclastogenesis but can also act in an autocrine manner through the IL-6R/gp130 complex. Signal transduction of IL-6R binding its ligand results in the up/down regulation of genes (X) responsible for osteoclast function. In addition, it has been suggested that infection of the osteoblast with RSV increases the levels of IL-11 which stimulates osteoclastogenesis. This regulation may also extend to other factors yet to be identified (Y). Elevation of specific cytokines in Paget's disease may be associated with the primary mechanism of pathogenesis or alternatively, arise as a secondary consequence of pathological activation. Adapted from ref. 120.

osteoblasts grown in vitro from pagetic and nonpagetic bone.[79] Using DNA amplification technology we determined a consistent pattern of expression, detecting transcripts for IL-1β, IL-3, IL-6, IL-8, TGFβ1, TGFβ2 and TGFβ3. It was not possible to identify mRNA for IL-1α, IL-2, IL-4, IL-5, IL-7, TNFβ or γ–interferon. This pattern of expression was identical whether the osteoblasts were grown from pagetic or nonpagetic bone, with the exception that TNFα was detected in some of the cultures of pagetic osteoblasts. The expression of TNFα by unstimulated pagetic osteoblasts in vitro may be secondary to the expression or overexpression of other factors, especially since TNFα production by cultured human osteoblasts is known to be stimulated by IL-1, GM-CSF and endotoxin.[83] Furthermore, recent studies have identified other cytokines (summarized in Table 5.1) produced by human osteoblasts, most of which have yet to be investigated in Paget's disease. Of particular interest are IGF-I and PTHrP[127] which we have recently shown to be differentially expressed by human osteoblasts and which may be related to abnormal remodeling.

A study by Ralston et al[128] investigated cytokine expression in biopsies of pagetic and nonpagetic bone. Transcripts detected included, IL-1α, IL-1β, IL-6, TNFα, TNFβ, BFGF, TGFβ and IGF-I. There were, however, no significant differences between the patterns of cytokine expression observed in pagetic and nonpagetic bone. It is interesting to note, however, that IL-6 transcripts could not be detected in 40% of patients with severe Paget's disease. In addition, 70% did not express IL-1α and 50% displayed no IL-1β mRNA. This study looked at the collective cytokine and growth factor expression of the cells in bone and, consequently, changes in cytokine production by specific cell(s) types pivotal to the abnormal remodeling observed in Paget's disease may have been masked or overlooked.

It is still not clear what is the underlying cause of Paget's disease, however, there is evidence that cytokines may mediate some of the bone cell function which leads to abnormal bone remodeling. IL-6, a cytokine which has been linked to postmenopausal osteoporosis, has been implicated in Paget's disease. Much of the in vitro evidence suggests that IL-6 acts on cells of the osteoclast lineage and unregulated overexpression of IL-6 results in hypercalcemia. Aberrant IL-6 expression in osteoporosis appears to "uncouple" the phases of bone resorption and formation. In Paget's disease where coupling of bone formation and resorption is preserved the elevated IL-6 levels may be a response to rather than a cause of increased osteoclastic activity.

Regulation of cytokine action in bone is complex and it is likely that quite subtle changes in cytokine and growth factor levels, undetected in previous studies, could significantly contribute to the pathogenesis of Paget's disease. To date, researchers have been selective in their choice of cytokines to investigate. There are many aspects of cytokine expression in Paget's disease which have not yet been considered including alternate splicing of mRNAs and gene polymorphisms which may predispose to disease. In addition, the cell context, i.e., receptor expression, extracellular matrix, second messenger and transcription factor status, undoubtedly influence the role cytokines can play in modulating cell function. Greater understanding of these processes will enable us to understand the regulatory mechanisms of normal bone remodeling and the abnormal bone turnover observed in Paget's disease.

REFERENCES

1. Gallagher JA. Human bone remodeling. In: R. Dulbecco ed. Encyclopedia of Human Biology, volume 1. Academic Press, Inc. 1991:811-824.
2. Parfitt AM. The coupling of bone formation to bone resorption: a critical analysis of the concept of its relevance to the pathogenesis of osteoporosis. Metab Bone Dis Rel Res 1982; 4:1-6.
3. Raisz LG. Local and systemic factors in the pathogenesis of osteoporosis. New Eng J Med 1988; 318:818-828.
4. MacDonald BR, Gowen M. Cytokines and bone. Brit J Rheum 1992; 31:149-155.
5. Nathan C, Sporn M. Cytokines in context. J Cell Biol 1991; 113:981-986.
6. Vaes G. Cellular biology and biochemical mechanism of bone resorption. Clin Ortho Rel Res 1988; 231:239-270.
7. Gowen M, Mundy GR. Actions of recombinant interleukin-1, interleukin 2 and interferon γ on bone resorption in vitro. J Immunol 1986; 136:2478.
8. Pfeilschifter J, Chenu C, Bird A, Mundy GR, Roodman GD. Interleukin-1 and tumour necrosis factor stimulate the formation

of human osteoclast-like cells in vitro. J Bone Miner Res 1989; 4:113-118.

9. Bertolini DR, Nedwin GE, Bringman TS, Smith DD, Mundy GR. Stimulation of bone resorption and inhibition of bone formation in vitro by human tumour necrosis factors. Nature 1986; 319:516-518.

10. Scheven BAA, Visser JWM, Nijweide PJ. In vitro osteoclast generation from different bone marrow fractions, including a highly enriched haemopoietic stem cell population. Nature 1986; 321:79-81.

11. Hattersley G, Chambers TJ. Effects of interleukin 3 and of granulocyte-macrophage and macrophage colony stimulating factors on osteoclast differentiation from mouse haemopoietic tissue. J Cell Physiol 1990; 142:201-209.

12. Shinar DM, Sato M, Rodan GA. The effect of haemopoietic growth factors on the generation of osteoclast-like cells in mouse bone marrow cultures. Endocrinology 1990; 126:1728-1735.

13. MacDonald BR, Mundy GR, Clark S, Wang EA, Kuehl TJ, Stanley ER, Roodman GD. Effects of human recombinant CSF-GM and highly purified CSF-1 on the formation of multinucleated cells with osteoclast characteristics in long-term bone marrow cultures. J Bone Miner Res 1986; 1:227-233.

14. Lorenzo JA, Sousa SL, Fonseca JM, Hock JM, Medlock ES. Colony-stimulating factors regulate the development of multinucleated osteoclasts from recently replicated cells in vitro. J Clin Invest 1987; 80:160-164.

15. Antonioloi Corboz V, Cecchini MG, Felix R, Fleisch H, van der Pluijm G, Löwik CWGM. Effect of macrophage colony-stimulating factor on in vitro osteoclast generation and bone resorption. Endocrinology 1992; 180:437-442.

16. Yoshida H, Hayashi SI, Kunisada T, Ogawa M, Nishikawa S, Okamura H, Sudo T, Shultz LD, Nishikawa S. The murine mutation osteoporosis is in the coding region of the macrophage colony stimulating factor gene. Nature 1990; 345:442-444.

17. Kurihara N, Bertolini D, Suda T, Akiyama Y, Roodman GD. Interleukin-6 stimulates osteoclast-like multinucleated cell formation in long-term human marrow cultures by inducing IL-1 release. J Immunol 1990; 144:426-430.

18. Löwik CWGM, van der Pluijm G, Bloys H, Hoekman K, Bijvoet OLM, Aarden LA, Papapoulos SE. Parathyroid hormon (PTH) and PTH-like protein (PLP) stimulate interleukin-6 production by osteogenic cells: a possible role of interleukin-6 in osteoclastogenesis. Biochem Biophys Res Commun 1989; 162:1546-1552.

19. Ishimi Y, Miyaura C, Jin CH, Akatsu T, Abe E, Nakamura Y,

Yamaguchi A, Yoshiki S, Matsuda T, Hirano T, Kishimoto T, Suda T. IL-6 is produced by osteoblasts and induces bone resorption. J Immunol 1990; 145:3297-3303.

20. Al-Humidan A, Ralston SH, Hughes DE, Chapman K, Aarden L, Graham R, Russell G, Gowen M. Interleukin-6 does not stimulate bone resorption in neonatal mouse calvariae. Jx Bone Miner Res 1991; 6:3-8.

21. Barton BE, Mayer R. IL-3 and IL-6 do not induce bone resorpion in vitro. Cytokine 1990; 2:217-220.

22. Black K, Garrett IR, Mundy GR. Chinese hamster ovarian cells transfected with the murine interleukin-6 gene cause hypercalcaemia as well as cachexia, leukocytosis and thrombocytosis in tumour-bearing nude mice. Endocrinology 1991; 128:2657-2659.

23. Reid IR, Lowe C, Cornish J, Skinner SJM, Hilton DJ, Wilson TA, Gearing DP, Martin TJ. Leukemia inhibitory factor—a novel bone-active cytokine. Endocrinology 1990; 126:1416-1420.

24. Tamura T, Udagawa N, Takahashi N, Miyaura C, Tanaka S, Yamada Y, Koishihara Y, Ohsugi Y, Kumaki K, Taga T, Kishimoto T, Suda T. Soluble interleukin-6 receptor triggers osteoclast formation by interleukin-6. Proc Natl Acad Sci USA 1993; 90: 11924-11928.

25. Girasole G, Passeri G, Jilka RL, Manolagas SC. Interleukin-11: a new cytokine critical for osteoclast development. J Clin Invest 1994; 93:1516-1524.

26. Takahashi N, Mundy GR, Kuehl TJ, Roodman GD. Osteoclast-like formation in foetal and newborn long term baboon marrow cultures is more sensitive to 1,25-dihydroxyvitamin D3 than adult long term marrow cultures. J Bone Miner. Res 1987; 2:311-317.

27. Shioi A, Teitelbaum SL, Ross FP, Welgus HG, Suzuki H, Ohara J, Lacey DL. Interleukin-4 inhibits murine osteoclast formation in vitro. J Cell Biochem 1991; 47:272-277.

28. Chenu C, Pfeilschifter J, Mundy GR, Roodman GD. Transforming growth actor β inhibits formation of osteoclast-like cells in long-term marrow cultures. Proc Natl Acad Sci 1988; 85:5683-5687.

29. Gowen M, Nedwin G, Mundy GR. Preferential inhibition of cytokine stimulated bone resorption by recombinant interferon gamma. J Bone Miner Res 1986; 1:469-474.

30. Watanabe K, Tanaka Y, Morimoto I, Yahata K, Zeki K, Fujihira T, Yamashita U, Eto S. Interleukin-4 as a potent inhibitor of bone resorption. Biochem Biophys Res Commun 1990; 172:1035-1041.

31. Thomson BM, Saklatvala J, Chambers TJ. Osteoblasts mediate interleukin-1 stimulation of resorption by rat osteoclasts. J Exp Med 1986; 164:104-112.

32. Thomson BM, Mundy GR, Chambers TJ. Tumour necrosis fac-

tors α and β induce osteoblastic cells to stimulate osteoclastic bone resorption. J Immunol 1987; 138:775.

33. Stashenko P, Dewhirst FE, Peros WJ, Kent RL, Ago JM. Synergistic interactions between interleukin-1, tumour necrosis factor and lymphotoxin in bone resorption. J Immunol 1987; 138:1464-1468.

34. Dewhirst FE, Ago JM, Peros WJ, Stashenko P. Synergism between parathyroid hormone and interleukin-1 in stimulating bone resorption in organ culture. J Bone Miner Res 1987; 2:127-134.

35. Rodan GA, Martin TJ. Role of osteoblasts in hormonal control of bone resorption: an hypothesis. Calcif Tissue Int 1981; 33:349-351.

36. Oreffo ROC, Mundy GR, Seyedin SM, Bonewald LF. Activation of the bone-derived latent TGF beta complex by isolated osteoclasts. Biochem Biophys Res Commun 1989; 158:817-823.

37. Fenton AJ, Martin TJ, Nicholson GC. Carboxyl-terminal parathyroid hormone-related protein inhibits bone resorption by isolated chicken osteoclasts. J Bone Miner Res 1994; 9:515-519.

38. Fenton AJ, Kemp BE, Hammonds RG, Mitchelhill K, Moseley JM, Martin TJ, Nicholson GC. A potent inhibitor of osteoclastic bone resorption within a highly conserved pentapeptide region of parathyroid hormone-related protein: PTHrP (107-111). Endocrinology 1991; 129:3424-3426.

39. Sone T, Kohno H, Kikuchi H, Ikeda T, Kasai R, Kikuchi Y, Takeuchi R, Konishi J, Shigeno C. Human parathyroid hormone-related peptide (107-111) does not inhibit bone resorption in neonatal mouse calvaria. Endocrinology 1992; 131:2742-2746.

40. Gowen M, Wood DD, Russell RGG. Stimulation of the proliferation of human bone cells in vitro by human monocyte products with interleukin-1-like activity. J Clin Invest 1985; 75:1223-1229.

41. Evans DB, Bunning RAD, Van Damme J, Russell RGG. Natural human IL-1β exhibits regulatory actions on human bone-derived cells in vitro. Biochem Biophys Res Commun 1989; 159:1242-1248.

42. Gowen M, MacDonald BR, Russell RGG. Actions of recombinant human-interferon and tumour necrosis factor-a on the proliferation and osteoblastic characteristics of human trabecular bone cells in vitro. Arthritis Rheum 1988; 31:1500-1507.

43. Rickard DJ, Gowen M, MacDonald BR. Proliferation responses to estradiol, IL-1α and TGFβ by cells expressing alkaline phosphatase in human osteoblast-like cell cultures. Calcif Tissue Int 1993; 52:227-233.

44. Canalis E, Centrella M, McCarthy T. Effects of basic fibroblast growth factor on bone formation in vitro. J Clin Invest 1988; 81:1572-1577.

45. Canalis E, McCarthy TL, Centrella M. Effects of platelet-derived growth factor on bone formation in vitro. J Cell Physiol 1989;

140:530-537.

46. McCarthy TL, Centrella M, Canalis E. Regulatory effects of IGF-I and IGF-II on bone collagen synthesis in rat calvarial cultures. Endocrinology 1988; 124:301-309.

47. Hock JM, Centrella M, Canalis E. IGF-I has independent effects on bone matrix formation and cell replication. Endocrinology 1988; 122:254-260.

48. Merriman HL, LaTour D, Linkhart TA, Mohan S, Baylink DJ, Strong DD. IGF-I and IGF-II induce *c-fos* in mouse osteoblastic cells. Calcif Tissue Int 1990; 46:258-262.

49. Scheven BA, Hamilton NJ, Fakkeldij TM, Duursma SA. Effects of rh IGF-I and IGF-II and GH on the growth of normal human osteoblast-like cells and human osteogenic sarcoma cells. Growth Regul 1989; 1:160-167.

50. Noda M, Camilliere JJ. In vivo stimulation of bone formation by transforming growth factor-beta. Endocrinology 1989; 124: 2991-2994.

51. Marcelli C, Yates AJ, Mundy GR. In vivo effects of human recombinant transforming growth factor-beta on bone turnover in normal mice. J Bone Miner Res 1990; 5:1087-1096.

52. Mackie EJ, Trechsel U. Stimulation of bone formation in vivo by transforming growth factor-beta—remodeling of woven bone and lack of inhibition by indomethacin. Bone 1990; 11:295-300.

53. Hock JM, Canalis E, Centrella M. Transforming growth factor-beta stimulates bone matrix apposition and bone cell replication in cultured fetal rat calvariae. Endocrinology 1990; 126:421-426.

54. Centrella M, McCarthy TL, Canalis E. Transforming growth factor-beta is a bifunctional regulator of replication and collagen synthesis in osteoblast-enriched cell cultures from fetal rat bone. J Biol Chem 1987; 262:2869-2874.

55. Bonewald LF, Kester MB, Schwartz Z, Swain L, Khare A, Johnson T, Leach R, Boyan B. Effects of combining transforming growth factor β and 1,25(OH)$_2$D$_3$ on differentiation of a human osteosarcoma (MG-63). J Biol Chem 1991; 267:8943-8949.

56. Noda M, Rodan G. Type B transforming growth factor inhibits proliferation and expression of alkaline phosphatase in murine osteoblast-like cells. Biochem Biophys Res Commun 1986; 140:56-65.

57. Luyten F, Cunningham N, Ma S, Muthukumaran N, Hammonds RG, Nevins WB, Wood W, Reddi AH. Purification and partial amino acid sequence of osteogenin, a protein initiating bone differentiation. J Biol Chem 1989; 264:13377-13380.

58. Wang EA, Rosen V, Cordes P, Hewick RM, Kriz MJ, Luxenberg DP, Sibley BS, Wozney JM. Purification and characterisation of other distinct bone-inducing factors. Proc Natl Acad Sci USA 1988;

85:9484-9488.

59. Wozney JM, Rosen V, Celeste AJ, Mitsock LM, Whitters MJ, Kriz RW, Hewick RM, Wang EA. Novel regulators of bone formation: molecular clones and activities. Science 1988; 242:1528-1534.

60. Dedhar S, Gaboury L, Galloway P, Eaves C. Human granulocyte-macrophage colony stimulating factor is a growth factor active on a variety of cell types of nonhemopoietic origin. Proc Natl Acad Sci USA 1988; 85:9253-9257.

61. Evans DB, Bunning RAD, Russell RGG. The effects of recombinant human granulocyte-macrophage colony-stimulating factor (rhGM-CSF) on human osteoblast-like cells. Biochem Biophys Res Commun 1989; 160:588-595.

62. Lowe C, Cornish J, Callon K, Martin TJ, Reid IR. Regulation of osteoblast proliferation by leukemia inhibitory factor. J Bone Miner Res 1991; 6:1277-1283.

63. Evans DB, Gerber B, Feyen JHM. Recombinant human leukemia inhibitory factor is mitogenic for human bone-derived osteoblast-like cells. Biochem Biophys Res Commun 1994; 199:220-226.

64. Metcalf D, Gearing DP. A fatal syndrome in mice engrafted with cells producing high levels of the leukemia inhibitory factor (LIF). Proc Natl Acad Sci USA 1989; 86:5948-5952.

65. Civitelli R, Martin TJ, Fausto A, Gunsta SL, Hruska KA, Avioli LV. Parathyroid hormone-related peptide transiently increases cytosolic calcium in osteoblast-like cells—comparison with parathyroid hormone. Endocrinology 1989; 125:1204-1210.

66. Rodan SB, Noda M, Wesolowski G, Rosenblastt M, Rodan GA. Comparison of postreceptor effects of 1-34 human hypercalcemic factor and 1-34 human parathyroid hormone in rat osteosarcoma cells. J Clin Invest 1988; 81:924-927.

67. Canalis E. Interleukin-1 has independent effects on deoxyribonucleic acid and collagen synthesis in cultures of rat calvariae. Endocrinology 1986; 118:74-81.

68. Canalis E. Effects of tumour necrosis factor on bone formation in vitro. Endocrinology 1987; 121:1596-1604.

69. Hanazawa S, Ohmori Y, Amano S, Hirose K, Miyoshi T, Kumegawa M, Kitano S. Human purified interleukin-1 inhibits DNA synthesis and cell growth of osteoblastic cell line (MC3T3-E1), but enhances alkaline phosphatase in the cells. FEBS Lett 1986; 203:279-284.

70. Dedhar S. Regulation of the expression of the cell adhesion receptors, integrins, by recombinant human interleukin-1β in human osteosarcoma cells: inhibition of cell proliferation and stimulation of alkaline phosphatase activity. J Cell Physiol 1989; 138:291-299.

71. Stashenko P, Dewhirst FE, Rooney ML, Desjardins LA, Heeley

JD. Interelukin-1β is a potent stimulator of bone formation in vitro. J Bone Miner Res 1987; 2:559-565.

72. Riancho JA, Zarrabeitia MT, Olmos JM, Amado JA, Gonzalez-Macias J. Effects of interleukin-4 on human osteoblast-like cells. Bone Miner 1993; 21:53-61.

73. Ueno K, Katayama T, Miyamoto T, Koshihara Y. Interleukin-4 enhances in vitro mineralization in human osteoblast-like cells. Biochem Biophys Res Commun 1992; 189:1521-1526.

74. Hughes FJ, Howells GL. Interleukin-11 inhibits bone fromation in vitro. Calcif Tissue Int 1993; 53:362-364.

75. Canalis E, Effect of IGF-I on DNA and protein synthesis in cultured rat alvariae. J Clin Invest 1980; 66:709-719.

76. Fournier B, Ferralli JM, Price PA, Schlaerppi JM. Comparison of the effects of IGF-I and IGF-II on the human osteosarcoma cell line OHS-4. J Endocrinol 1993; 136:173-180.

77. Chenu C, Valentin-Opran A, Chavassieux P, Saez S, Meunier PJ, Delmas PD. IGF-I hormonal regulation by GH and 1,25 vitamin D and activity on human osteoblast-like cells in short term cultures. Bone 1990; 11:81-86.

78. Harris SE, Bonewald LF, Harris MA, Sabatini M, Dallas S, Feng JQ, Gnosh-Choudhury N, Wozney J, Mundy GR. Effects of transforming growth factor β on bone nodule formation and expression of bone morphogenetic protein 2, osteoclacin, osteopontin, alkaline phosphatase, and type I collagen mRNA in long-term cultures of fetal rat calvarial cells. J Bone Miner Res 1994; 9:855-863.

79. Birch MA, Ginty AF, Walsh CA, Fraser WD, Gallagher JA, Bilbe G. PCR detection of cytokines in normal human and pagetic osteoblast-like cells. J Bone Miner Res 1993; 8:1155-1162.

80. Chaudhary LR, Spelsberg TC, Riggs BL. Production of various cytokines by human osteoblast-like cells in response to interleukin-1β and tumour necrosis factor-α: lack of regulation by 17β-estradiol. Endocrinology 1992; 130:2528-2534.

81. Elias JA, Tang W, Horowitz MC. Cytokine and hormonal stimulation of human osteosarcoma interleukin-11 production. Endocrinology 1995; 136:489-498.

82. Keeting PE, Rifas L, Harris SA, Colvard DS, Spelsberg TC, Peck WA, Riggs BL. Evidence for interleukin-1β production by cultured normal human osteoblast-like cells. J Bone Miner Res 1991; 6:827-833.

83. Gowen M, Chapman K, Littlewood AJ, Hughes DE, Evans DB, Russell RGG. Production of tumour necrosis factor by human osteoblasts is modulated by other cytokines, but not by osteotropic hormones. Endocrinology 1990; 126:1250-1255.

84. Littlewood AJ, Russell J, Harvey G, Hughes DE, Russell RGG.

The modulation of the expression of IL-6 and its receptor in human osteoblasts in vitro. Endocrinology 1991; 129:1513-1520.

85. Oursler MJ, Cortese C, Keeting P, Anderson MA, Bonde SK, Riggs BL, Spelsberg TC. Modulation of transforming growth factor-β production in normal human osteoblast-like cells by 17β-estradiol and parathyroid hormone. Endocrinology 1991; 129:3313-3320.

86. McCarthy TL, Centrella M, Canalis E. PTH enhances the transcript and polypeptide levels of IGF-I in osteoblast-enriched cultures from fetal rat bone. Endocrinology 1989; 124:1247-1253.

87. Gray TK, Mohan S, Linkhart TA, Baylink DJ. Estradiol stimulates in vitro the secretion of IGFs by the clonal osteoblastic cell line UMR 106. Biochem Biophys Res Commun 1989; 158:407-412.

88. Canalis E, Pash J, Gabbitas B, Rydziel S, Varghese S. Growth factors regulate the synthesis of insulin-like growth factor-I in bone cell cultures. Endocrinology 1993; 133:33-38.

89. Kishimoto T, Taga T, Akira S. Cytokine signal transduction. Cell 1994; 76:253-262.

90. Nathan C, Sporn M. Cytokines in context. J Cell Biol 1991; 113:981-986.

91. Rogelj S, Klagsbrun M, Atzmon R, Kurokawa M, Haimovitz A, Fuks Z, Vlodavsky I. Basic fibroblast growth factor is an extracellular matrix component required for supporting the proliferation of vascular endothelial cells and the differentiation of PC12 cells. J Cell Biol 1989; 109:823-831.

92. Gordon MY, Rile GP, Watt SM, Greaves MF. Compartmentalization of a haemopoietic growth factor (GM-CSF) by glycosaminoglycans in the bone marrow microenvironment. Nature 1987; 326:403-405.

93. Roberts R, Gallagher J, Spooncer E, Allen TD, Bloomfield F, Dexter TM. Heparan sulphate bound growth factors: a mechanism for stromal cell mediated haemopoiesis. Nature 1988; 332:376-378.

94. Yamaguchi Y, Mann DM, Ruoslahti E. Negative regulation of transforming growth factor-β by the proteoglycan decorin. Nature 1990; 346:281-284.

95. Dike LE, Farmer SR. Cell adhesion induces expression of growth-associated genes in suspension-arrested fibroblasts. Proc Natl Acad Sci USA 1988; 85:6792-6796.

96. Hankey DP, Hughes AE, Mollan RAB, Nicholas RM. Extracellular protein secretion of cultured normal and pagetic osteoblasts. Electrophoresis 1993; 14:644-649.

97. Garrett IR, Durie BGM, Nedwin GE, Gillespie A, Bringman T, Sabatini M, Bertolini DR, Mundy GR. Production of lymphotoxin, a bone resorbing cytokine, by cultured human myeloma cells. N Engl J Med 1987; 526-532.

98. Cozzolino F, Torcia M, Aldinucci D, Rubartelli A, Miliani A, Shaw AR, Lansdorp PM, Diguglielmo R. Production of interleukin-1 by bone marrow myeloma cells. Blood 1989; 74:380-387.

99. Bataille R, Jourdan M, Zhang Xue-Guang, Klein B. Serum levels of interleukin-6, a potent myeloma cell growth factor as a reflection of disease severity in plasma cell dyscrasias. J Clin Invest 1989; 84:2008-2011.

100. Mundy GR. Hypercalcemia in hematologic malignancies and in solid tumors associated with extensive localised bone destruction. In: Favus MJ ed. Primer on the metabolic bone diseases and disorders of mineral metabolism. Raven Press, New York, 1993:173-176.

101. Nakamoto T, Chang C, Li A, Chodak GW. Basic fibroblast growth factor in human prostate cancer cells. Cancer Res 1992; 52:571-577.

102. Massachusetts General Hospital: Case records of the Massachusetts General Hospital (Case 27461). N Engl J Med 1941; 225:789-791.

103. Martin TJ, Moseley JM, Gillespie MT. Parathyroid hormone-related protein: biochemistry and molecular biology. Crit Rev Biochem Mol Biol 1991; 26:377-395.

104. Bouizar Z, Spyratos F, Deytieux S, De Vernejoul M, Julliene A. Polymerase chain reaction analysis of parathyroid hormone-related protein gene expression in breast cancer patients and occurence in bone metastases. Cancer Res 1993; 53:5076-5078.

105. Luparello C, Ginty AF, Gallagher JA, Pucci-Minafra S. Transforming growth factor β- 1,2 and 3, urokinase and parathyroid hormone-related peptide expression in 8701-BC breast cancer cells and clones. Differentiation 1993; 55:73-80.

106. Nouri AME, Panayi GS, Goodman SM. Cytokines and the chronic inflammation of rheumatic disease I. The presence of interleukin-1 in synovial fluids. Clin Exp Immunol 1984; 55:295-300.

107. Houssiau FA, Devogelaer JP, Van Damme J, Nagant de Deuxchaisnes C, van Nicj J. Interleukin-6 in synovial fluid and serum of patients with rheumatoid arthritis and other inflammatory arthritides. Arthritis Rheum 1988; 31:784-788.

108. Saxne T, Palladino MA, Heinegard D, Talal N, Wolheim FA. Detection of tumour necrosis factor α but not tumour necrosis factor β in rheumatoid arthritis synovial fluid and serum. Arthritis Rheum 1988; 31:1041-1045.

109. Firestein GS, Xu WD, Townsend K, Broide D, Alvarogracia J, Glasebrook A, Zvaifler NJ. Cytokines in chronic inflammatory arthritis I. Failure to detect T cell lymphokines (interleukin 2 and interleukin 3) and presence of macrophage colony stimulating factor and a novel mast cell growth factor in rheumatoid synovitis. J Exp Med 1988; 168:1573-1586.

110. Pacifici R, Rifas L, Teitelbaum S, Slatopolsky E, McCraken R,

Bergfeld M, Lee W, Avioli LV, Peck WA. Spontaneous release of interleukin-1 from human blood monocytes relects bone formation in idiopathic osteoporosis. Proc Natl Acad Sci USA 1987; 84:4616-4620.

111. Ralston SH, Russell RGG, Gowen M. Estrogen inhibits release of tumour necrosis factor from peripheral blood mononuclear cells in postmenopausal women. J Bone Miner Res 1990; 5:983-988.

112. Pacifici R, Brown C, Puscheck E, Friedrich E, Slatopolsky E, Maggio D, McCracken R, Aviolo LV. Effect of surgical menopause and estrogen replacement on cytokine release from human blood mononuclear cells. Proc Natl Acad Sci USA 1991; 88:5134-5138.

113. Girasole G, Jilka RL, Passeri G, Boswell S, Boder G, Williams DC, Manolagas SC. 17β-estradiol inhibits interleukin-6 production by bone-marrow derived stromal cells and osteoblasts in vitro: a potential mechanism for the anti-osteoporotic effect of estrogens. J Clin Invest 1992; 89:883-891.

114. Jilka RL, Hangoc G, Girasole G, Passeri G, Williams DC, Abrams JS, Boyce B, Broxmeyer H, Manolagas SC. Increased osteoclast development after oestrogen loss: mediation by interleukin-6. Science 1992; 257:88-91.

115. Bennett AE, Wahner HW, Riggs BL, Hintz RL. Insulin-like growth factors I and II: aging and bone density in women. J Clin Endocrinol Metab 1984; 59:701-704.

116. Donahue LR, Hunter SJ, Sherblom AP, Rosen CJ. Age-related changes in serum IGFBPs in women. J Clin Endocrinol Metab 1990; 71:575-579.

117. Johansson AG, Burman P, Westermark K, Ljunghall S. Bone mineral density in acquired GH deficiency correlates with circulating levels of IGF-I. J Inten Med 1992; 232:447-452.

118. Pioli G, Girasole G, Pedrazzoni M, Sansoni P, Erroi A, Davoli L, Ciotti G, Mantovani A, Passeri M. Spontaneous release of interleukin-1 (IL-1) from medullary mononuclear cells of pagetic subjects. Calcif Tissue Int 1989; 45:257-259.

119. Roodman GD, Kurihara N, Ohsaki Y, Kukita T, Hosking D, Demulder A, Singer FS. Interleukin-6: a potential autocrine/paracrine factor in Paget's disease of bone. J Clin Invest 1992; 89:46-52.

120. Hoyland JA, Freemont AJ, Sharpe PT. Interleukin-6, IL-6 receptor, and IL-6 nuclear factor gene expression in Paget's disease. J Bone Miner Res 1994; 9:75-80.

121. Dodds RA, Merry K, Littlewood A, Gowen M. Expression of mRNA for IL1β, IL6 and TGFβ1 in developing human bone and cartilage. J Histochem Cytochem 1994; 42:733-744.

122. Kiyokawa T, Yamaguchi K, Takaya M, Takahashi K, Watanabe T,

Matsumoto T, Lee SY, Takatsuki K. Hypercalcemia and osteoclast proliferation in adult T-cell leukemia. Cancer 1987; 59:1187.

123. Shirakawa F, Yamashita U, Tanaka Y, Watanabe K, Sato K, Hiratake J, Fujihira T, Oda O, Eto S. Production of bone-resorbing activity corresponding to interleukin-1α by adult T-cell leukaemia cells in humans. Cancer Res 1988; 48:4284.

124. Watanabe T, Yamaguchi K, Takatsuki K, Osame M, Yoshida M. Constitutive expression of parathyroid hormone-related protein gene in human T cell leukemia virus type 1 (HTLV-1) carriers and adult T cell leukemia patients that can be *trans*-activated by HTLV-1 *tax* gene. J Exp Med 1990; 172:759-765.

125. Bieberich CJ, King CM, Tinkle BT, Jay G. A transgenic model of transactivation by the tax protein of HTLV-1. Virology 1993; 196:309-318.

126. Elias J, Zheng T, Einarsson O, Landry M, Trow T, Rebert N, Panuska J. Epithelial interleukin-11: regulation by cytokines, respiratory syncytial virus and retinoic acid. J Biol Chem (in press).

127. Walsh CA, Birch MA, Fraser WD, Lawton R, Dorgan J, Walsh S, Sansom D, Beresford JN, Gallagher JA. Expression and secretion of parathyroid hormone-related protein by human bone-derived cells in vitro: effects of glucocorticoids. J Bone Miner Res 1995; 10:17-25.

128. Ralston SH, Hoey SA, Gallacher SJ, Adamson BB, Boyle IT. Cytokine and growth factor expression in Paget's disease: analysis by reverse-transcription/polymerase chain reaction. Brit J Rheumat 1994; 33:620-625.

129. Middleton J, Arnott N, Walsh S, Beresford JN. Osteoblasts and osteoclasts in adult human osteophyte tissue express the mRNAs for insulin-like growth factors I and II and the type 1 IGF receptor. Bone 1995; 16:287-293.

======= CHAPTER 6 =======

A MOLECULAR MODEL
OF PAGET'S DISEASE

Paul T. Sharpe

INTRODUCTION

For a disease that was first described over 100 years ago that may well affect something like five million people worldwide there is embarrassingly little known about the molecular biology of Paget's disease. If Paget's disease was an inherited, genetic disease, it would certainly have attracted far greater attention from researchers. The treatment of Paget's disease is also something of an enigma; excellent treatments have been developed, principally involving bisphosphonate drugs, with little idea of the molecular biology or biochemistry of the disease, without any idea of the cause of the disease and with little idea of the mechanisms of action of the drugs themselves.

The possible causes of Paget's disease are discussed throughout this volume and the current status of the actions of bisphosphonates are discussed in chapter 7. The aim of this chapter is to summarize the most recent work on identifying molecules whose expression is altered in Paget's disease and how this information can be assembled to produce a molecular model for the altered molecular biology of pagetic bone cells. Such a model is an important step forward not only because it establishes links with a possible viral

The Molecular Biology of Paget's Disease, edited by Paul T. Sharpe.
© 1996 R.G. Landes Company.

cause, but because it provides a framework for experimentation, i.e., it makes testable predictions.

THE PAGETIC OSTEOCLAST

Paget's is a disease of the osteoclast. The primary lesion is believed to occur in osteoclasts to locally increase bone resorption. What follows is a breakdown in the normal coupling between osteoclasts (resorption) and osteoblasts (formation) resulting in the uncontrolled formation of bone characteristic of Paget's disease. The morphology of pagetic bone that is seen in sections of biopsies is thus caused by uncontrolled osteoblast activity, but what sets the whole process in motion is a lesion in osteoclasts. At the cellular level the most striking characteristic of pagetic bone is the increase in the number, size and nuclear content of osteoclasts. Whether the increase in bone resorption results from the increase in osteoclast numbers alone or whether pagetic osteoclasts are also more active is not clear.[3] What is clear is that whatever the cause of Paget's disease, pagetic osteoclasts are the primary target cell and an understanding of how they differ from nonpagetic osteoclasts is an important goal. With this aim in mind we and others have set out to characterize molecules (genes) whose expression is altered in pagetic osteoclasts. Because pagetic bone is a complex mixture of different cells, the development of which we know very little, we have used techniques that detect the presence of molecules as they are found in vivo. The technique we have mainly used is in situ hybridization to detect expression of genes coding for proteins known to have important roles in bone cell function.

Table 6.1 shows a summary of the in situ hybridization results obtained so far. Sections were used from at least six different Paget's patients and controls for active but nonpagetic osteoclasts were sections of osteophytes from femoral heads isolated during hip replacement surgery. A qualitative assessment of the level of gene expression was obtained by estimation of silver grains over cells in random fields of view and scored on a scale of three from an average for the six patients. The results of these experiments highlight some interesting expression patterns of important genes. The first point of note is that none of the genes tested were found to be expressed at a reduced level in pagetic bone cells. This would imply that both pagetic osteoblasts and osteoclasts are more active in a general sense with respect to gene transcription and probably protein synthesis than equivalent cells in osteophytes. A second general point of interest is that several genes such as BMP2 and BMP4 whose expression has been assumed to be osteoblast specific were also found to be expressed in osteoclasts.

Among these differentially expressed genes there are two which are specifically and highly upregulated in pagetic osteoclasts and are thus of particular interest These are the cytokine interleukin-6 (IL-6) and the proto-oncogene *c-FOS*. *IL-6* expression was detected in marrow osteoclast precursors by in situ hybridization and this expression was accompanied by an increase in the level of IL-6 protein in serum of Paget's patients.[3] Based on these observations David Roodman proposed that IL-6 acts as a autocrine/paracrine factor promoting the recruitment of osteoclast precursors from the marrow to the bone surface.[1] IL-6 thus seems to have a potentially important role in pagetic osteoclasts. However, detection of IL-6 protein and/or expression in a cell does not necessarily mean that the cell is a target cell. IL-6 protein is secreted and acts on cells only after binding to receptors. Thus an essential prerequisite for IL-6 action in pagetic osteoclasts is the presence of IL-6 receptors. In situ hybridization has revealed that the IL-6 receptor gene

Table 6.1. Relative levels of gene expression in osteoblasts and osteoclasts from pagetic bone and nonpagetic bone.

	Paget's		Non-Paget's	
	OB	OC	OB	OC
TGFβ1	+	+	+	+
TGFα	+	+	+	−
IL-6	+ +	+ + +	+	−
IL-6R	+	+ +	+	+
BMP 2	+ +	+ +	+	+
BMP 3	+ +	+ +	+	+
BMP 4	+	+	−	−
NF-IL-6	+	+	+	+
c-FOS	+	+ + +	−	−

Scoring is based on grain-counts over remodeling selected cells following in situ hybridization with ^{35}S-labeled RNA.

(*IL-6R*) is also expressed in pagetic osteoclasts as is the transcription factor, NF-IL-6 through which IL-6 exerts its direct effects on gene transcription.[2] Pagetic osteoclasts are thus fully equipped to produce and respond to IL-6.

The proto-oncogene *c-FOS* has a long association with bone that goes back to the original discovery of *vFOS* in osteosarcomas.[3] c-FOS is able to heterodimerize with Jun protein family members and bind to AP-1 cis-acting sites found in a large number of genes. *c-Fos* was also one of the first genes whose expression was shown to be spatially regulated in developing skeletal tissues during embryogenesis.[4,5] Despite this association with bone, until relatively recently c-FOS was ignored by most bone researchers largely because of its wide-ranging expression in many cell types other than bone. The essential role of c-FOS in bone is now unequivocal due almost entirely to the work of Erwin Wagner and his group in Vienna (reviewed in ref 6). *c-FOS* gene expression has been shown to be specifically upregulated in pagetic osteoclasts and is accompanied by an increase in c-FOS protein.[7] This is particularly relevant in light of the role of c-FOS in bone since upregulation of c-FOS in transgenic mice results in osteoblast transformation and osteosarcoma formation,[8,9] whereas inactivation of c-Fos causes the bone remodeling disease osteopetrosis due to the absence of osteoclasts.[10,11]

Considering IL-6 and c-FOS upregulation in pagetic osteoclasts, it is significant that IL-6 has been shown to upregulate *c-FOS* expression in B cell differentiation and pheochromocytoma cells.[12,13] It is possible therefore that IL-6 upregulates *c-FOS* in pagetic osteoclasts, thus providing a molecular link between osteoclast precursor recruitment and activation (Fig. 6.1). Furthermore, it has been established using knock-out mice that c-FOS is essential for osteoclast differentiation from marrow precursors.[10,11] Thus, elevated c-FOS expression in marrow precursors might result in increased numbers of osteoclasts.

If IL-6 upregulates *c-FOS* expression in pagetic osteoclasts, how might IL-6 expression itself be upregulated? One intriguing possibility is that the ubiquitous transcription factor, NF-κB, is activated in pagetic osteoclasts. NF-κB is present in all cells in an inactive form coupled to an inhibitor, Iκβ.[14] NF-κB is activated by

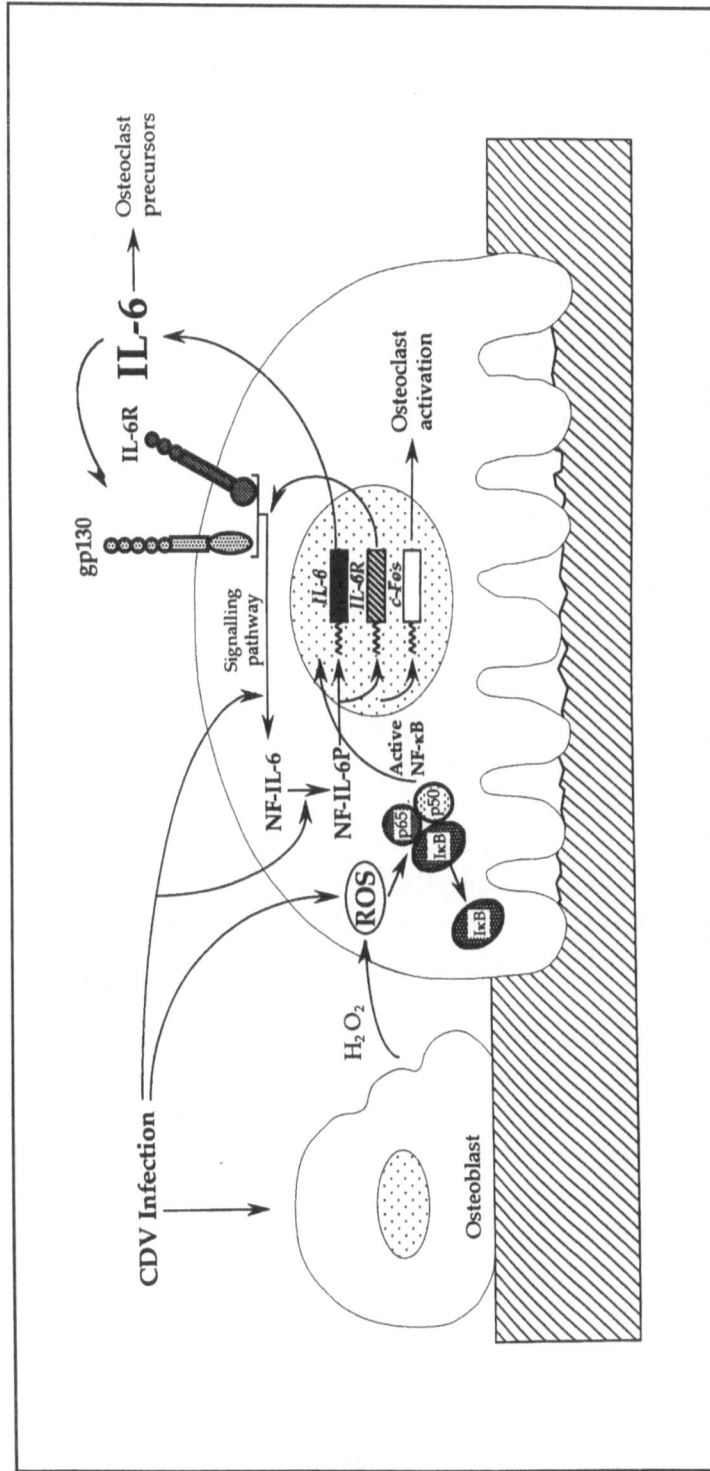

Fig. 6.1. Viral infection of osteoclasts and/or osteoblasts stimulates the production of ROS which in turn activate NF-κB by removal of the inhibitory IκB. NF-κB then up-regulates IL-6 gene expression. Alternatively, the virus might act directly on the signaling pathway to NF-IL-6, or in activation of NF-IL-6, possibly by phosphorylation. Increased activity of NF-IL-6 then up-regulates IL-6 and IL-6R gene expression. Increased IL-6 upregulates c-Fos gene expression. Increased IL-6R levels result in an increased response to IL-6, increased levels of c-Fos stimulate osteoclast activity, and IL-6 in the serum promotes osteoclast precursor recruitment and/or fusion. Adapted from Hoyland et al 1994.

dissociation of Iκβ to release functional NF-κB in the cytoplasm which then enters the nucleus. Although at present we do not know whether active NF-κB is present in pagetic osteoclasts, its presence is implied from experiments where expression of an HTLV-1 LTR *Tax* transgene in mice results in bone lesions with characteristics of Paget's disease.[15] *Tax* is known to stimulate NF-κB activity together with a number of cytokines including IL-6, and furthermore, in mouse epidermal cells, NF-κB is known to induce IL-6 production.[16-18] Although Tax is not expressed in pagetic osteoclasts and is unlikely to have any role in Paget's disease, it demonstrates that activation of NF-κB in osteoclasts could be an important feature of Paget's disease (Fig. 6.1). If this is correct and we assume that NF-κB is specifically activated in pagetic osteoclasts, which in turn upregulates *IL-6* gene expression, then we must search for ways in which NF-κB might be activated.

Many agents have been shown to activate NF-κB in different cells but three which require consideration in the context of Paget's disease are viruses, reactive oxygen species and IL-1. Reactive oxygen species (ROS) have attracted much attention in many fields of biology. ROS are known to stimulate osteoclast formation and activity.[19,20] It has been shown that hydrogen peroxide produced from osteoblasts can be taken up by osteoclasts, and thus it seems likely that osteoclasts do harbor ROS.[21] To complete the link between NF-κB activation, IL-6 and c-FOS, ROS would have to be specifically produced in pagetic osteoclasts as opposed to other active but nonpagetic osteoclasts. It is at this point that this putative pathway leads back to the possible etiological agent, namely paramyxovirus infection (chapter 4) since ROS have been demonstrated in cultured brain cells infected with the canine paramyxovirus, canine distemper virus (CDV).[22] Thus, a pathway of connective events can be assembled from paramyxovirus infection of osteoclasts generating ROS which activate NF-κB resulting in upregulation of IL-6 expression which is secreted from the cells where it acts as a paracrine agent, stimulating precursor recruitment, and in an autocrine loop via receptors on the osteoclast surface, to activate c-FOS expression. Finally, c-FOS acts to activate osteoclast function or, if expressed in early osteoclast precursors, directs more precursors to become osteoclasts. A potent stimulator of bone resorption, IL-1 has recently been shown to be upregulated NF-κB in osteoclasts, providing a direct link between bone resorption and NF-κB.[23]

The final part of this pathway is perhaps the most interesting because it provides a possible explanation for three of the most tantalizing and unexplained features of Paget's disease: namely, its focal nature; its persistence in an inactive form over many years; and reoccurrence at the original site of infection following withdrawal of treatment. If the primary lesion in Paget's disease is paramyxoviral infection of a small, localized population of stem cells in the marrow that occurs in a patient's youth, the virus could persist indefinitely in the stem cells and remain localized to a particular part of the marrow. Sometime in later life, Paget's disease could be triggered by the upregulation of c-FOS in this localized group of precursor cells, leading to an increase in osteoclasts at the bone surface. The stimulus for this trigger might be accumulation of ROS above a threshold level. The virus remains in the stem cells as they differentiate into osteoclasts where the presence of its fusion and hemagglutinin proteins promotes cell fusion, generating large osteoclasts. The focal nature of the disease is explained by the fact that the virus is only ever present in a very small number of precursor cells that are localized to a particular place in the marrow. These few precursor cells are assumed only to populate one or two sites in the skeleton and that in some way these sites are repopulated by stem cells from the same marrow location following withdrawal of treatment.

PAGETIC OSTEOBLASTS

The role of osteoblasts is important for the phenotype but assumed to be secondary in the etiology. Nevertheless, the detection of paramyxoviruses in pagetic osteoblasts remains an important unanswered question. Continuing the theme of viral infection of precursor cells in the marrow, it might be possible that a small population of osteoblast stromal cell precursors are infected, although this seems unlikely since these infected precursors would have to end up at precisely the same sites in the skeleton as the infected osteoclast precursors. It should be added, however, that it has never been convincingly shown that osteoblasts from unaffected sites in Paget's patients are negative for paramyxoviruses, so there remains the possibility that all osteoblasts in Paget's patients harbor virus. If there is virus in osteoblasts then it is open to question whether it plays a functional part in the pagetic phenotype. If ROS can be induced in osteoclasts by viral infection then there is no reason to expect anything different in osteoblasts. Since it has been shown that hydrogen peroxide from osteoblasts can affect osteoclasts, the viral infection of osteoblasts may have a secondary affect on osteoclasts through the release of hydrogen peroxide. One problem with this is that you would expect the same NF-κB /IL-6/c-FOS pathway to be activated in pagetic osteoblasts, but this clearly does not happen, at least to the same extent as in osteoclasts. The low level of IL-6 receptor RNA detected in pagetic osteoblasts indicates that these cells are poorly responsive to IL-6 signals, so it is possible that the pathway is incomplete.

My own personal view is that the infection of osteoblasts (by whatever route) probably plays no role in establishment of the pagetic lesion but may have a role once it is established by producing hydrogen peroxide which amplifies the IL-6/c-FOS pathway already underway in osteoclasts.

A MULTIFACTORIAL DISEASE?

An obvious consequence of this model of Paget's disease is that it suggests that Paget's may be considered to be a multifactorial disease. Thus, although viral infection may be a primary cause of the disease in many instances, there could be an underlying genetic basis (predisposition?) that explains the instances of familial incidence. Mutations or polymorphisms in different genes in the pathways predicted in the model (e.g., c-FOS, IL-6, NF-κB) could all conceivably result in the same phenotype. In this respect Paget's disease might resemble multifactorial disorders such as cleft palate. Congenital cleft palate is a malformation whose basis is multifactorial in that it can be caused by mutations in different genes. This is most clearly illustrated by gene knock-out studies in transgenic mice where disruption of a number of different developmentally regulated genes have all resulted in a cleft palate phenotype. Cleft palate also has familial forms such as X-linked cleft palate.

TESTABLE PREDICTIONS

There are a number of predictions of this molecular model of Paget's disease, some of which are more easily testable than others. Work in my group is underway to investigate some of the more obvious features of the model such as NF-κB and ROS activities in pagetic osteoclasts. A particularly interesting recent publication showed that in rheumatoid arthritis, bisphosphonates act through inhibiting ROS.[22] Another direction we are taking is using the model as a basis for producing features of Paget's disease in transgenic animals. Thus, we have recently produced transgenic mice which ectopically express c-FOS specifically in osteoclasts and are currently examining the resulting phenotype.

REFERENCES

1. Roodman GD, Kurihara N, Ohsaki Y, Kukita A, Hosking D, Demulder A, Singer FR. Interleukin-6: A potential autocrine/paracrine factor in Paget's disease of bone. J Clin Invest 1992; 89:46-52.
2. Hoyland JA, Freemont AJ, Sharpe PT. Interleukin-6 (IL-6), IL-6 receptor and IL-6 nuclear factor gene expression in Paget's disease. J Bone Min Res 1993; 9:75-80.
3. Finkel MP, Biskis BO, Tinkins PB. Virus of osteosarcomas in mice. Science 1966; 151:698-702.
4. Caubert JF, Bernaudin JF. Expression of the c-Fos proto-oncogene in bone, cartilage and tooth forming tissues during mouse development. Biol Cell 1988; 64:101-104.
5. Dony C, Gruss P. Proto-oncogene c-Fos expression in growth regions of fetal bone and mesodermal web tissue. Nature 1987; 328:711-714.
6. Grigoriadis AE, Wang Z-q, Wagner EF. Fos and bone cell development: Lessons from a nuclear oncogene. Trends Genet 1995; 11:436-441
7. Hoyland JA, Sharpe PT. Up-regulation of c-Fos protooncogene expression in pagetic osteoclasts. J Bone Min Res 1994; 9: 1191-1194.
8. Ruther U, Garber C, Komitowski D, Muller R, Wagner EF. Deregulated c-Fos expression interferes with normal bone development in transgenic mice. Nature 1987; 325:412-416.
9. Wang Z-Q, Liang J, Schellander K, Wagner EF, Grigoriadis AE. c-Fos induced osteosarcoma formation in transgenic mice: Cooperativity with c-Jun and the role of endogenous c-Fos. Cancer Res 55; 6244-6251.

10. Grigoriadis AE, Wang Z-Q, Cecchini MG, Hofstetter W, Felix R, Fleisch HA, Wagner EF. c-Fos is a key regulator of osteoclast/macrophage lineage determination and bone remodelling. Science 1994; 266:443-448.

11. Wang Z-Q, Ovitt C, Grigoriadis AE, Möhle-Steinlein U, Rüther U, Wagner EF. Bone and haematopoietic defects in mice lacking c-fos. Nature 1992; 360; 741-745.

12. Korholz D, Gerdan S, Enczmann J, Zessack N, Burdach S. Interleukin-6 induced differentiation of a human B cell line into IgM secreting plasma cells is mediated by c-Fos. Eur J Immunol 1992; 22:607-610.

13. Metz R, Ziff E. cAMP stimulates the C/EBP-related transcription factor rNFIL-6 to trans-locate to the nucleus and induce c-Fos transcription. Genes Dev 1991; 5:1754-1766.

14. Baeuerie P, Baltimore D. A 65-KD subunit of active NF-κB is required for inhibition of NF-κB by Iκβ. Genes Dev 1989; 3: 1689-1698.

15. Ruddle NH, LiC-B, Horne WC, Santiago P, Troiano N, Jay G, Horowitz M, Baron R. Mice transgenic for HTLV-1 LTR-tax exhibit tax expression in bone, skeletal alterations and high bone turnover. Virology 1993; 197:196-204.

16. Libermann TA, Baltimore D. Activation of interleukin-6 gene expression through the NF-κB transcription factor. Mol Cell Biol 1990; 10:2327-2334.

17. Shimizu H, Mitomo K, Watanabe T, Okamoto S, Yamamoto K-I. Involvement of a NF-κB-like transcription factor in the activation of the interleukin-6 gene by inflammatory lymphokines. Mol and Cell Biol 1990; 10:561-568.

18. Brach MA, de Vos S, Arnold C, Gruss H-J, Mertelsmann R, Hermann F. Leukotriene B_4 transcriptionally activates interleukin-6 expression involving NF-κB and NF-IL6. Euro J Immunol 1992; 22:2705-2711.

19. Garrett IR, Boyce BF, Oreffo ROC, Bonewald L, Poser J, Mundy GR. Oxygen derived free radicals stimulate osteoclastic bone resorption in rodent bone in vitro and in vivo. J Clin Invest 1990; 85:632-639.

20. Suda N, Testa NG, Allen TD, Onions D, Jarrett O. Effects of hydrocortisone on osteoclasts generated in cat bone marrow cultures. Calcified Tissue International 1993; 35:82-86.

21. Bax BE, Alam ASMT, Banerji B, Bax CMR, Bevis PJR, Stevens CR, Moonga BS, Blake DR, Zaida M. Stimulation of osteoclastic bone resorption by hydrogen peroxide. Biochem Res Comm 1992; · 183:1153-1158.

22. Bürge T, Griot C, Vandevelde M, Peterham E. Antiviral antibod-

ies stimulate production of reactive oxygen species in cultured canine brain cells infected with canine distemper virus. J Viol 1989; 63:2790-2797.

23. Jimi E, Ikebe T, Takahashi N, Hirata M, Suda T, Koga T. Interleukin-1α activates NF-κB in osteoclasts. Bone 1996; 17:582.

CHAPTER 7

MECHANISMS OF ACTION OF BISPHOSPHONATES AS INHIBITORS OF BONE RESORPTION

Michael J. Rogers and R. Graham G. Russell

ANTIMINERALIZATION AND ANTIRESORPTIVE PROPERTIES OF BISPHOSPHONATES

More than 25 years have passed since it was first recognized that bisphosphonates could inhibit bone resorption.[1-3] During this time, bisphosphonates have become the treatment of choice for a variety of bone diseases in which excessive osteoclast activity is an important pathological feature, including Paget's disease of bone,[4] metastatic and osteolytic bone disease, hypercalcemia of malignancy[5] and, more recently, postmenopausal and other forms of osteoporosis (Fig. 7.1).[6,7] However, despite the widespread clinical use of bisphosphonates, their exact mechanism of action, at both the cellular and molecular level, have not been clearly identified.

The discovery of bisphosphonates as anti-resorptive agents was partly serendipitous. In the 1960s, attempts were being made to identify agents that could regulate the deposition of bone mineral (hydroxyapatite) during calcification and that could therefore be

The Molecular Biology of Paget's Disease, edited by Paul T. Sharpe.
© 1996 R.G. Landes Company.

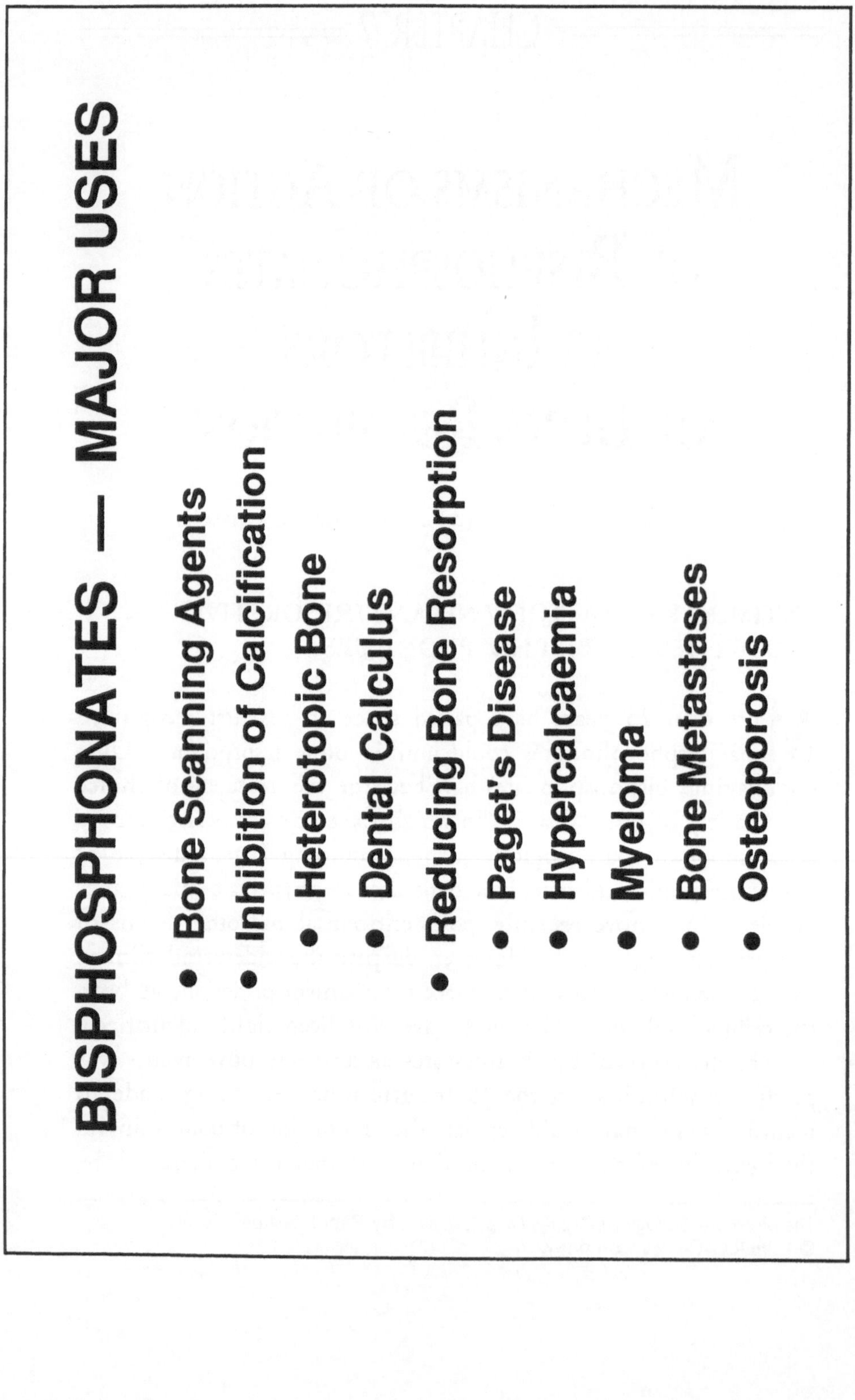

BISPHOSPHONATES — MAJOR USES

- Bone Scanning Agents
- Inhibition of Calcification
  - Heterotopic Bone
  - Dental Calculus
- Reducing Bone Resorption
  - Paget's Disease
  - Hypercalcaemia
  - Myeloma
  - Bone Metastases
  - Osteoporosis

Fig. 7.1. Major clinical uses of bisphosphonates.

useful in the prevention of heterotopic calcification. Inorganic pyrophosphate, a natural by-product of many biosynthetic reactions in the body, was identified as one such agent that was present in serum and urine and could prevent calcification by binding to newly-forming crystals of hydroxyapatite.[8] It was therefore postulated that locally-high concentrations of pyrophosphate could prevent bone mineralization, and that extracellular enzymes that hydrolyze pyrophosphate, such as alkaline phosphatase, could regulate the concentration of pyrophosphate in the bone microenvironment.[9,10] Orally-administered pyrophosphate and polyphosphates were unable to inhibit ectopic calcification in laboratory animals, due to the hydrolysis of pyrophosphate in the gastrointestinal tract. Bisphosphonates, formerly also known as diphosphonates and which are nonhydrolyzable analogs of pyrophosphate that contain stable P-C-P bonds rather than labile P-O-P bonds, were therefore studied as compounds that might also have the anti-mineralization properties of pyrophosphate but that would be resistant to hydrolysis.[11]

Bisphosphonic acid   Pyrophosphoric acid

Indeed, like pyrophosphate, bisphosphonates did appear to have high affinity for bone mineral and could prevent calcification in vitro, but, unlike pyrophosphate, were also able to prevent pathological calcification when given orally to rats in vivo.[12,13] The ability of the bisphosphonates to bind to bone mineral, preventing both crystal growth and dissolution, was enhanced when the R^1 side chain (attached to the geminal carbon atom of the P-C-P group) was a hydroxyl group (as in etidronate, or 1-hydroxyethylidene-1,1-bisphosphonate) rather than a halogen atom such as chlorine (as in clodronate, or dichloromethylene-1,1-bisphosphonate).[14] The presence of this hydroxyl group imparts a higher affinity for calcium (and thus bone mineral) due to the ability of the bisphosphonates to chelate calcium ions by tridentate rather than bidentate binding.[15] Thus, bisphosphonates appear to prevent calcification by a physicochemical mechanism, acting as crystal poisons after adsorption to bone surfaces.[16]

In addition to their antimineralization properties, bis-phosphonates were found to have the novel property of being able to inhibit the dissolution of hydroxyapatite crystals, and this led to studies of whether they might also inhibit bone resorption. Numerous studies using a variety of experimental systems showed unequivocally that they were able to inhibit osteoclast-mediated bone resorption, both in organ cultures of bone in vitro and in thyroparathyroidectomized rats treated with parathyroid hormone to stimulate bone resorption in vivo.[14] It was clearly demonstrated that inhibition of bone resorption required a P-C-P group and could not be achieved with monophosphonates, e.g., pentane monophosphonate,[3] or with P-C-C-P or P-N-P compounds (Fig. 7.2). The anti-resorptive effect could not be accounted for simply by adsorption of bisphosphonates to bone mineral and prevention of hydroxyapatite dissolution, since clodronate, although having less affinity for hydroxyapatite than etidronate, was a more potent anti-resorptive agent. Thus, it appeared that the mechanism by which bisphosphonates inhibited bone resorption was different from the mechanism by which they inhibited bone mineralization. The identification of more potent anti-resorptive bisphosphonates also supports this view, since alterations to the structure of the R^2 side chain, but without altering R^1 (and hence affinity for bone mineral), have a dramatic effect on anti-resorptive potency.[17,18] In particular, the presence of a basic primary nitrogen atom within an alkyl chain (as in pamidronate and alendronate) increases potency by 10- to 100-fold relative to etidronate and clodronate,[19] whilst derivatives of these compounds that contain a tertiary nitrogen (such as ibandronate and olpadronate)[20,21] are even more potent. Amongst the most potent anti-resorptive bisphosphonates are those containing a nitrogen atom within a heterocyclic ring (as in EB1053, risedronate and zoledronate), which are up to 10,000-fold more potent than etidronate (Fig. 7.3, 7.4).[22-24]

Although the structure of the R^2 side chain largely determines the potency of the bisphosphonate (Fig. 7.2), alterations to the

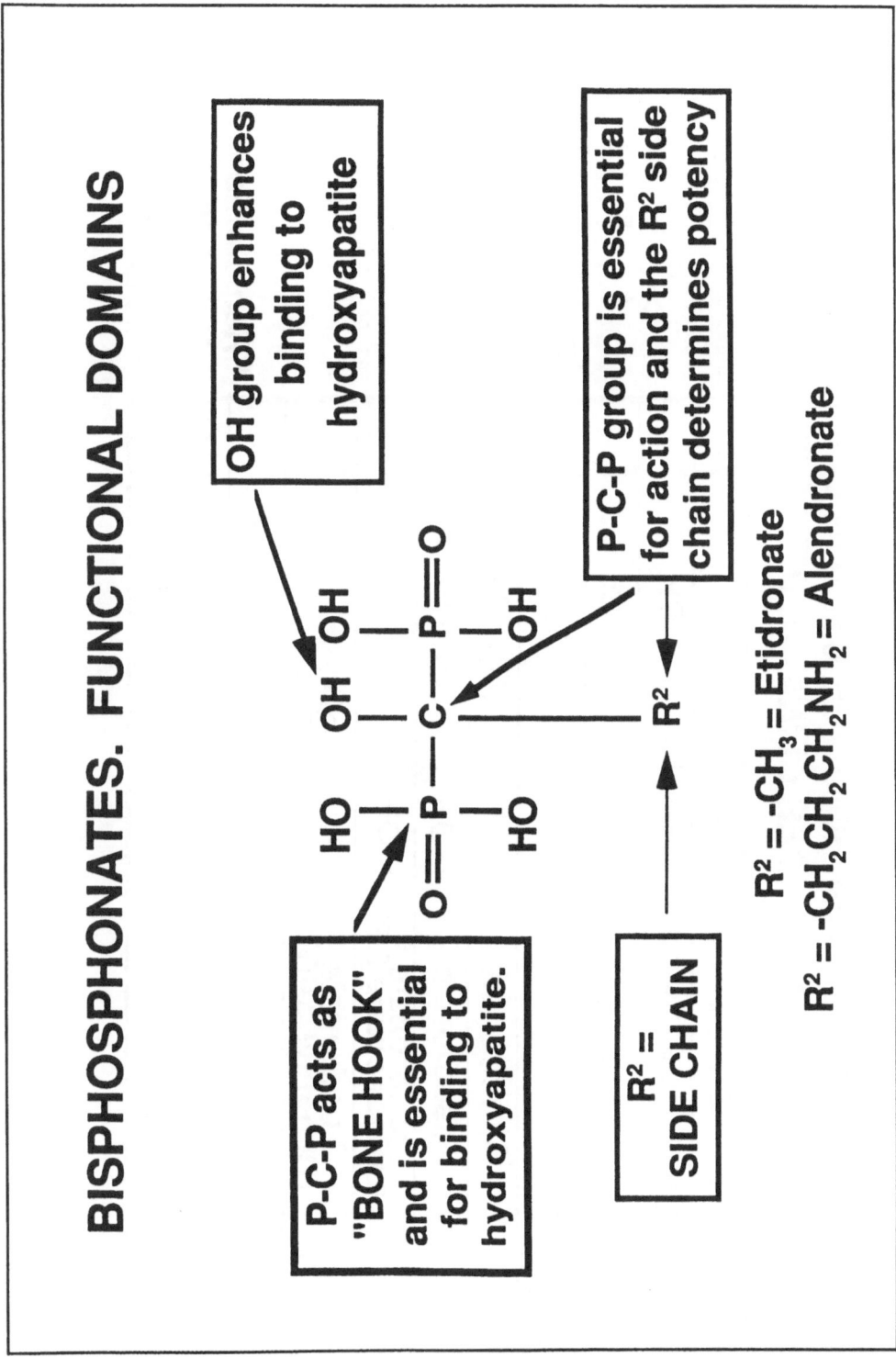

Fig. 7.2. Structure of a bone-active bisphosphonate to show functional domains.

SOME BISPHOSPHONATES USED IN CLINICAL STUDIES AND UNDER CLINICAL DEVELOPMENT

$$O=\overset{OH}{\underset{OH}{P}}-\overset{R'}{\underset{R''}{C}}-\overset{OH}{\underset{OH}{P}}=O$$

BISPHOSPHONATE	R'	R''
Etidronate (EHDP)*	OH	CH_3
Clodronate (Cl$_2$MDP)*	Cl	Cl
Pamidronate (APD)*	OH	$CH_2CH_2NH_2$
Alendronate*	OH	$(CH_2)_3NH_2$
Risedronate	OH	CH_2-3-pyridine
Neridronate (Aminohexane DP)	OH	$(CH_2)_5NH_2$
Olpadronate (Dimethyl APD)	OH	$CH_2CH_2N(CH_3)_2$
Tiludronate	H	CH_2-S-phenyl-Cl
Ibandronate	OH	$CH_2CH_2N(CH_3)$(pentyl)
Zoledronate	OH	CH_2 -(imidazole)

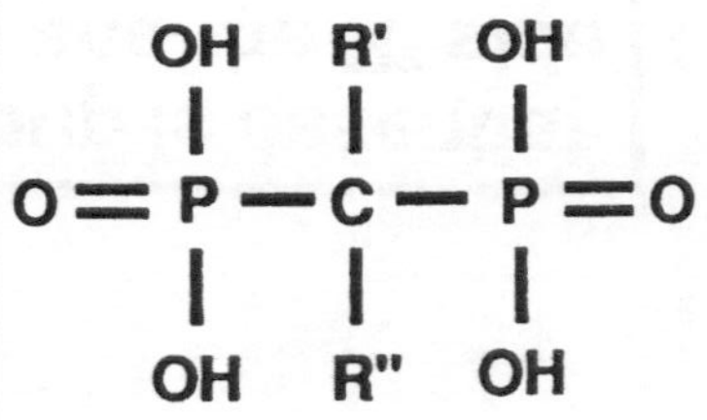

* Registrations approved in some countries

Fig. 7.3. Structures of bisphosphonates used in clinical studies.

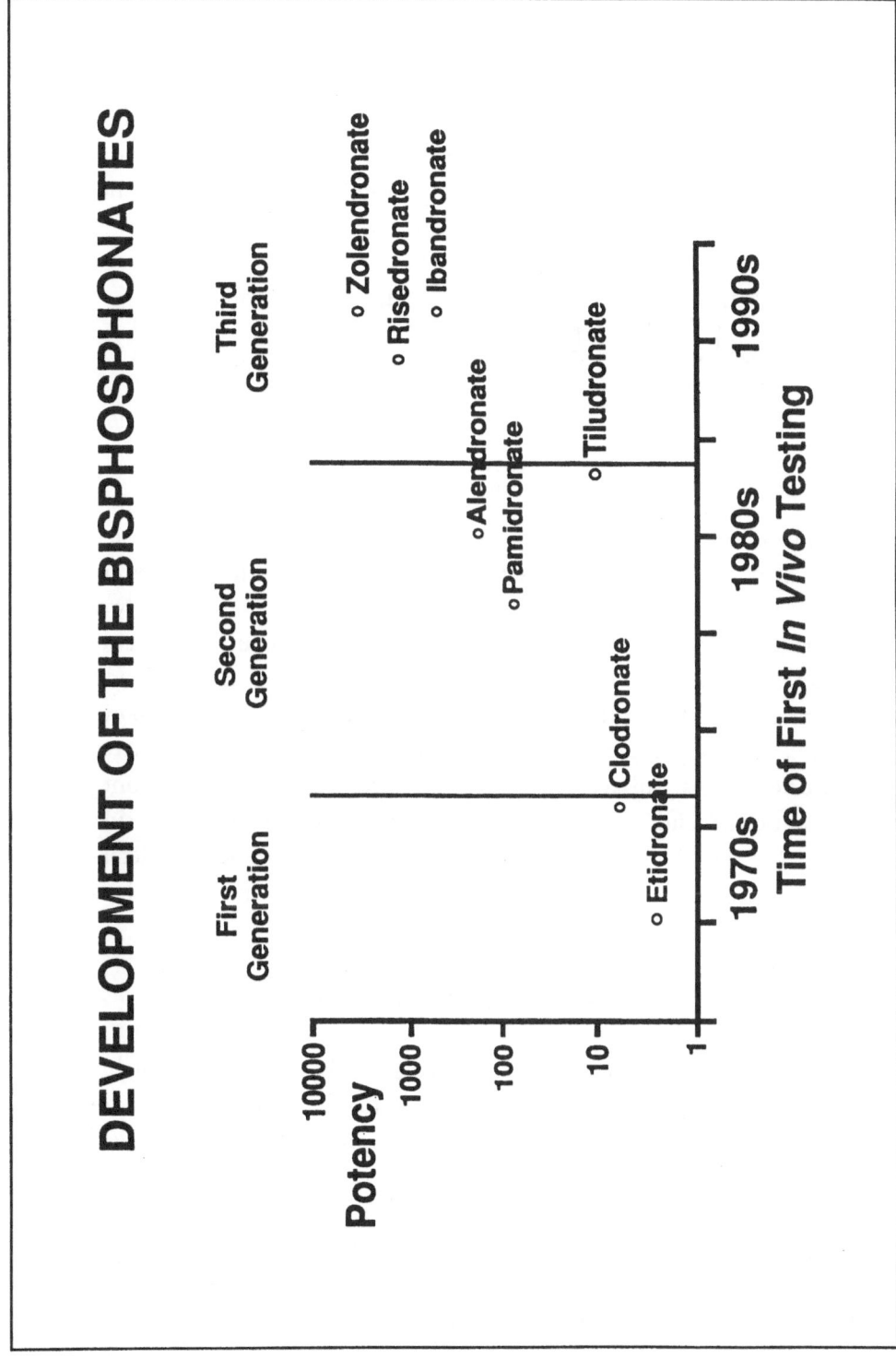

Fig. 7.4. Increasing potency of newly described bisphosphonates of clinical relevance.

phosphonate groups that affect affinity for hydroxyapatite can also affect anti-resorptive potency. For example, replacement of one of the phosphonate hydroxyl groups with a methyl group (to form a phosphonomethylphosphinate) markedly reduces both bone affinity and anti-resorptive potency. Alteration of both phosphonate groups in this way (to form a bisphosphinate) leads to loss of bone affinity and loss of anti-resorptive activity.[16,18]

bisphosphonate   phosphono-methylphosphinate   bisphosphinate

These studies of the relationships between bisphosphonate structure and anti-resorptive potency suggest that the ability of bisphosphonates to inhibit bone resorption is dependent on two separate properties of the bisphosphonate molecule. The two phosphonate groups, together with a hydroxyl group as the R^1 side chain, impart high affinity for bone mineral and act as a "bone hook", allowing rapid and efficient targeting of bisphosphonates to bone mineral surfaces (Fig. 7.2).[25] Once localized within bone, the structure and three dimensional conformation of the R^2 side chain (as well as the phosphonate groups in the molecule) determine the biological activity of the molecule, a property which is still the subject of intense study.

DIRECT EFFECTS OF BISPHOSPHONATES ON OSTEOCLASTS

Bone resorption in vivo is mediated solely by osteoclasts—highly specialized, multinucleate cells that are formed by the fusion of hemopoietic, unicellular precursors in the bone marrow (reviewed by Mundy and Roodman, and by Zaidi et al).[26,27] Resorption of bone takes place following the adhesion of osteoclasts to the bone surface, a process that involves the recognition of extracellular matrix proteins by membrane-bound receptors such as the $\alpha_v\beta_3$ integrin.[28,29] Localized sites of focal adhesion (podosomes) together with polarization of the cytoskeleton into characteristic rings of microfilaments,[30] results in the formation a tight seal between the plasma membrane of the osteoclast and the bone surface. Resorption of bone then occurs in the region beneath the osteoclast, into which is secreted a battery of proteolytic enzymes and which is acidified by the action of a vacuolar-type, plasma membrane $H^+ATPase$.[31-33] The low pH of this microenvironment beneath the osteoclast causes dissolution of the hydroxyapatite bone mineral, while the digestion of the extracellular bone matrix is brought about by the action of the proteolytic enzymes.[34]

Due to their high affinity for calcium, bisphosphonates are rapidly cleared from the circulation in vivo (or from the extracellular medium in vitro in the presence of bone) and adsorb to bone mineral.[35] Thus, concentrations of bisphosphonate in solution that were originally very low could give rise to very much higher local concentrations after adsorption then release from bone (Fig. 7.5). Osteoclasts, of all the cell types in bone, are the most likely to be exposed to the highest concentrations of free, nonmineral-bound bisphosphonate, due to the release of the bisphosphonate from bone mineral in the low pH environment beneath osteoclasts resorbing bisphosphonate-coated bone. For example, Sato et al have estimated that pharmacological doses of alendronate that inhibit bone resorption in vivo could give rise to local concentrations as high as 1mM alendronate in the resorption space beneath an osteoclast.[36]

The most obvious route by which bisphosphonates could inhibit bone resorption is therefore by a direct effect on bone resorbing osteoclasts that have released bisphosphonate during the resorption process.[37] Since osteoclasts are highly endocytic, bisphosphonate present in the resorption space is likely to be internalized by endocytosis. Indeed, osteoclasts in vivo have been shown to internalize radiolabeled alendronate into numerous intracellular vacuoles, with evidence of the bisphosphonate also being present in other subcellular compartments, including the cytoplasm, mitochondriae and nuclei.[36] Following internalization, bisphosphonates could inhibit resorption either through a toxic effect that would lead to osteoclast cell death or by interfering with one or more of the processes vital for bone resorption, such as the activity of the proton-pumping ATPase (Fig. 7.5).

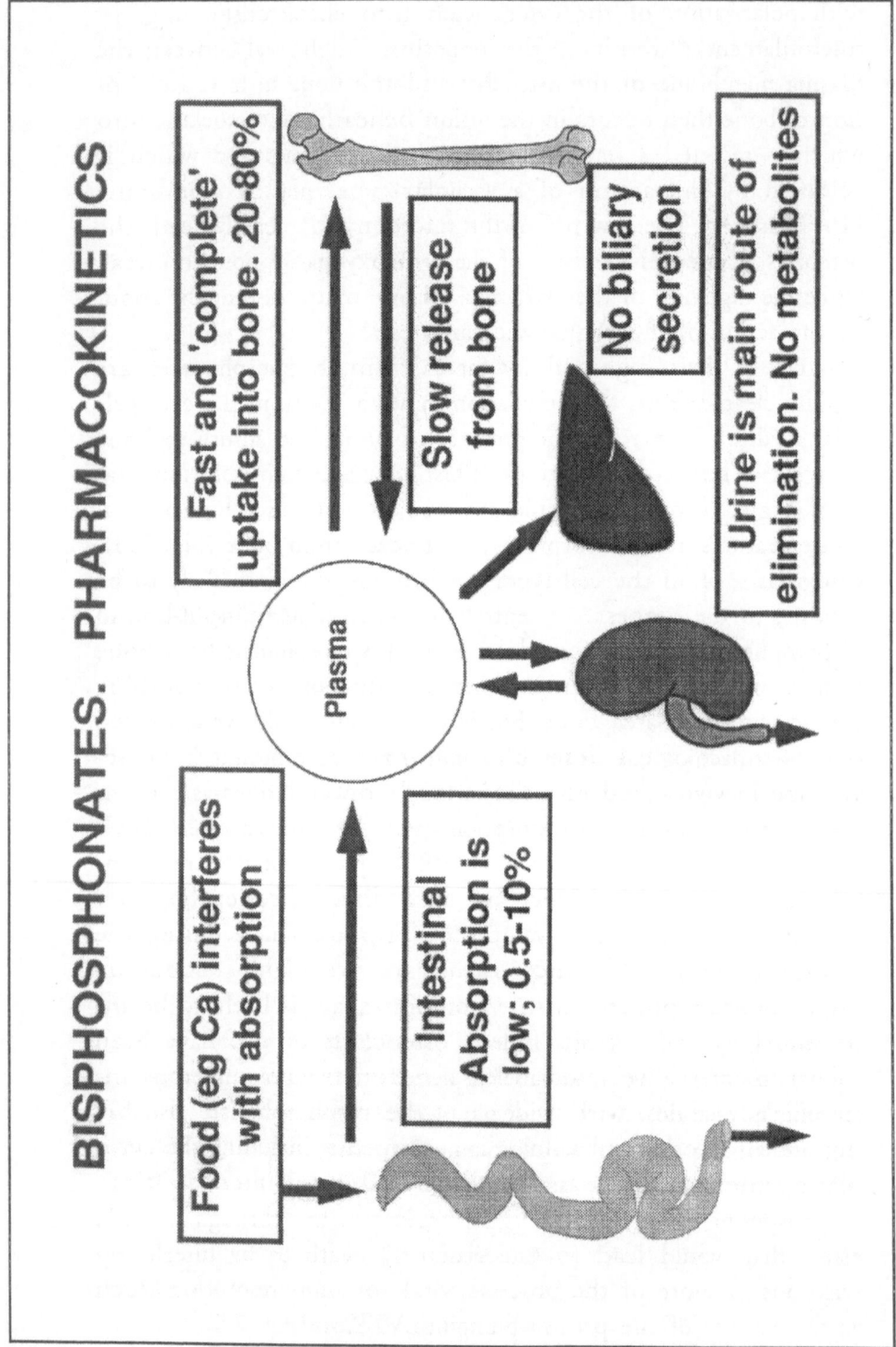

Fig. 7.5. Diagram to show the pharmacokinetic behavior of bisphosphonates in clinical practice.

METABOLIC INHIBITION AND CELL DEATH

Observations made by Schenk et al,[3] Rowe et al[38-40] and others[41] suggested that bisphosphonates could be toxic to mature osteoclasts, since degenerative changes in cell morphology, such as nuclear pyknosis and fragmentation, were observed in osteoclasts both in vitro and in histological sections from rats treated with clodronate and etidronate. In retrospect, these morphological features are suggestive of apoptosis (or "cellular suicide"), a form of cell death that can be distinguished from necrosis on the basis of characteristic changes in morphology (cell and nuclear condensation, chromatin condensation and nuclear fragmentation) as well as biochemical events, including internucleosomal cleavage of DNA following the activation of an endonuclease.[42,43] More recently, Hughes et al[44] have demonstrated that bisphosphonates do indeed induce apoptosis in mouse osteoclasts both in vitro and in vivo. Following treatment with clodronate, pamidronate and risedronate, the characteristic morphological features of apoptosis could be identified in both isolated osteoclasts in vitro and in osteoclasts in histological sections. In addition, DNA cleavage could also be demonstrated in osteoclasts in tissue sections by using the TUNEL assay, a technique that detects apoptotic cells following the enzyme-catalyzed incorporation of a histochemical label into DNA strand breaks caused by endonuclease activity. Selander et al have also shown that apoptotic cell death can be induced in isolated osteoclasts by clodronate.[45]

It is becoming clear that apoptosis plays a major role in the regulation of bone turnover and remodeling, and that some endogenous factors that inhibit bone resorption (such as estrogen and TGFβ) promote osteoclast apoptosis, whereas factors that stimulate resorption (such as interleukin-1) prevent osteoclast apoptosis.[44,46,47] Thus, a major mechanism by which bisphosphonates inhibit bone resorption could be to directly trigger premature osteoclast cell death by apoptosis. This could be achieved by causing changes in the level of certain second messengers such as intracellular calcium, an effect that has been observed in some osteoclasts in vitro by Sato et al.[36] Alternatively, osteoclast apoptosis may occur as a consequence of some other effect. Carano et al demonstrated that bisphosphonates inhibit bone resorption in vitro by a direct effect on osteoclasts, possibly due to an inhibitory effect on cell metabolism.[48] Since bisphosphonates resemble pyrophosphate and, presumably, also other naturally-occurring phosphorylated compounds, it is likely that bisphosphonates interact with at least some metabolic enzymes, an event that could lead to cell death or simply to an incapacity to fulfill the substantial metabolic requirements (of energy and macromolecular synthesis) for resorbing bone. For instance, several bisphosphonates have been found to inhibit protein synthesis in isolated osteoclasts and chondrocytes,[48,49] while clodronate and etidronate can inhibit glycolysis and reduce lactate production in isolated cells and calvaria.[50-52] Several amino-containing bisphosphonates, but not clodronate or etidronate, are also inhibitors of squalene synthetase or other enzymes involved in the biosynthesis of sterols.[53] There are many such reports of effects of bisphosphonates on metabolic enzymes or on biosynthetic reactions (reviewed by Felix[54]). However, many studies have not been carried out or repeated using a wide range of bisphosphonates. Hence, there is little data to suggest a correlation between the ability of bisphosphonates to inhibit these enzymes and the potency of the bisphosphonates as inhibitors of bone resorption. Furthermore, some of these biochemical effects may be indirect, and a consequence of some other effect of the bisphosphonates.

INHIBITION OF OSTEOCLAST FUNCTION

Bisphosphonates could also inhibit the ability of osteoclasts to resorb bone by a variety of mechanisms which, in turn, may eventually lead to osteoclast cell death. A characteristic morphological feature of bisphosphonate-treated osteoclasts is the lack of a ruffled border, the region of convoluted, apical plasma membrane facing the resorption cavity.[3,55-58] The formation of the ruffled border, as well as the sealing zone of the osteoclast, is dependent on the arrangement of actin, vinculin and other cytoskeletal elements. Not surprisingly, disruption of this cytoskeletal arrangement and loss of the ruffled border prevents bone resorption, as has been demonstrated with colchicine treatment[39] and in osteopetrotic, *c-src*-deficient mice.[59] Sato et al[36] and Murakami et al[58] have demonstrated that alendronate and tiludronate disrupt the formation of the cytoskeletal actin ring of polarized, resorbing osteoclasts. Microinjection of tiludronate into isolated osteoclasts also produced the same result, indicating that the effect was brought about by an intracellular, rather than extracellular, mechanism.[58] While it is possible that bisphosphonates could interact with cytoskeletal proteins directly, disruption of the cytoskeleton could also be brought about indirectly by inhibition of protein kinases or phosphatases that regulate cytoskeletal structure, such as c-src. In a recent report, Schmidt et al have shown that alendronate and etidronate can inhibit several protein tyrosine phosphatases (such as PTP_e and PTP_ε), without affecting serine threonine phosphatase PP-1.[60] Although the kinase activity of c-src itself was not inhibited by alendronate, bisphosphonates did inhibit the ability of protein tyrosine phosphatases, especially PTPmeg1, to dephosphorylate a c-src peptide substrate. Given the importance of c-src for osteoclastic resorption, and that another bisphosphonate (tiludronate) has also been shown to increase the level of protein phosphorylation in osteoclasts by inhibiting protein phosphatase activity,[61] inhibition of protein dephosphorylation and hence signal transduction in osteoclasts may be a further mechanism by which bisphosphonates inhibit the ability of osteoclasts to resorb bone.

Resorption of bone and degradation of the extracellular matrix occurs in the resorption cavity by means of lowered pH and the action of proteolytic enzymes.[34] Since the action of a proton-pumping ATPase is vital for acidification of the resorption cavity beneath active osteoclasts, inhibition of this H^+ ATPase, either directly or indirectly, would be expected to inhibit bone resorption. One bisphosphonate, tiludronate, has recently been claimed to have a direct and potent inhibitory effect on the proton-pumping activity of inside-out vesicles derived from osteoclast plasma membranes.[62] By contrast, Carano et al found that etidronate, pamidronate and clodronate could inhibit vacuolar acidification in intact osteoclasts, but that the effect was not due to direct inhibition of the H^+ATPase, since the bisphosphonates did not affect acidification of membrane-derived vesicles.[48] Similarly, Zimolo et al found that alendronate did not affect the vacuolar H^+ ATPase of osteoclast-derived vesicles.[63] Carano et al concluded that inhibition of proton pumping activity was a consequence of some form of metabolic inhibition, possibly protein synthesis.[48] It is possible that osteoclasts can extrude protons by more than one mechanism. Although osteoclasts have been shown to possess a bafilomycin A$_1$-sensitive H$^+$ATPase,[32] Zimolo et al demonstrated that osteoclasts also contain a bafilomycin A$_1$-insensitive, sodium-independent H$^+$ATPase, the activity of which is present in whole osteoclasts but not in membrane-derived vesicles, and is induced when osteoclasts are cultured on bone.[63] Furthermore, the activity of this H$^+$ATPase was not present when osteoclasts were cultured on alendronate- or etidronate-treated bone, one explanation being that the bisphosphonates may prevent protein or membrane trafficking and hence the insertion of the ATPase into the plasma membrane. Such an effect could be the result of an alteration in the osteoclast cytoskeleton, as discussed above. Bisphosphonates have also been shown to inhibit directly the activity of certain hydrolytic enzymes, e.g., phosphatases, and to prevent the release of lysosomal enzymes in mouse calvaria.[64,65] These effects may therefore also contribute to the overall inhibition of bone resorption.

INHIBITION OF OSTEOCLAST DEVELOPMENT

Continuous resorption of bone in vivo is totally dependent on the generation of new osteoclasts by the fusion of mononucleate, hemopoietic precursors. One possible mechanism by which bisphosphonates could inhibit bone resorption is therefore via osteoclast precursors, in addition to effects on mature osteoclasts.[66]

Hughes et al have shown that bisphosphonates can inhibit dose-dependently the formation of osteoclast-like cells in vitro in long-term cultures of human marrow stimulated with 1,25 dihydroxy-vitamin D$_3$.[67] Furthermore, the order of potency of the five bisphosphonates studied for inhibiting osteoclast formation in this system (risedronate>pamidronate>neridronate>clodronate>etidronate) matches the order of potency of the bisphosphonates for inhibiting bone resorption in vivo. In ex vivo organ culture also, bisphosphonates have been shown to affect the generation of mature osteoclasts. Boonekamp et al found that resorption of metacarpals from 17 day-old fetal mice could be inhibited by much lower concentrations of pamidronate than those required to block resorption of radii from the same mice.[68] Since the metacarpals from mice of this age do not contain mature osteoclasts, resorption is dependent on the generation of new osteoclasts from precursors present in the periosteum. By contrast, mature osteoclasts are already present in the radii. The conclusion reached was that the amino-containing bisphosphonates such as pamidronate could inhibit bone resorption by two mechanisms, involving an effect on mature osteoclasts at higher concentrations of bisphosphonate, and an effect on the generation of osteoclasts that could occur with lower concentrations. The less potent bisphosphonates such as clodronate and etidronate appeared to act predominantly on mature osteoclasts.

In order to examine further the mechanism by which bisphosphonates could prevent osteoclast development, Boonekamp et al used fetal liver as a source of hemopoietic osteoclast precursors.[68] Pretreatment of osteoclast-free mouse metacarpals, stripped to remove osteoclast precursors, with bisphosphonate for 24 hours before coculture with liver resulted in complete inhibition of resorption of the bone explant. This appeared to reflect an effect on the formation of new osteoclasts, since resorption of bone under these conditions is entirely dependent on the migration of osteoclast precursors from the liver to the bone, followed by the maturation then fusion of the precursors to form mature osteoclasts. Since pretreatment of the fetal liver with bisphosphonates did not affect subsequent resorption, and since separation of the liver from the bone explant by a membrane filter did not affect the migration of the osteoclast precursors to the mineral surface,[68,69] it was suggested that bisphosphonates may interfere with the recognition of some matrix factor which is essential for the transformation, maturation or fusion of the precursors to form mature osteoclasts. The mechanism by which this occurs has not yet been identified, although the fact that the relative potencies of the bisphosphonates in the coculture system match the relative in vivo antiresorptive potencies of the bisphosphonates[21,70] adds support to the view that this may be an important mechanism of action of some bisphosphonates. It is perhaps more likely that in vivo bisphosphonates affect both osteoclast precursors and mature osteoclasts, depending on the particular bisphosphonate and the extent to which the cells are exposed to it.

EFFECTS OF BISPHOSPHONATES ON OTHER BONE CELLS

OSTEOBLASTS

Although osteoclasts are perhaps the most likely cells to experience prolonged exposure to bisphosphonates in vivo, this does not exclude the possibility that bisphosphonates may also affect other cells in bone. In vitro, bisphosphonates have been shown to inhibit the proliferation of connective tissue cells (fibroblasts, chondrocytes and calvarial cells)[51,71] and osteoblast-like ROS 17/2.8 cells.[72] In the latter case, the order of potency for inhibition of cell proliferation (risedronate>alendronate>pamidronate>clodronate>etidronate) matches the order of anti-resorptive potency for the five bisphosphonates studied.

Since osteoblasts closely regulate the resorptive activity of osteoclasts, it is possible that bisphosphonates could inhibit osteoclasts via osteoblasts, either by preventing the osteoblasts from secreting an osteoclast-stimulatory factor, or by stimulating the osteoblasts to produce an osteoclast-inhibitory factor. Evidence for the latter effect has been reported by Sahni and co-workers, who found that treatment of rat osteoblast-like CRP 10/30 cells (which are potent stimulators of osteoclast activity) with 0.1 µM ibandronate or 1 µM clodronate resulted in inhibition of resorption when these cells were consequently cocultured for 24 hours with osteoclasts.[73] Surprisingly, the inhibitory effect could be achieved following even a short (5 minute) exposure of the CRP 10/30 cells to bisphosphonates, and also occurred when conditioned medium from the bisphosphonate-treated CRP 10/30 cells was added to osteoclast cultures. Subsequently, it was reported that the inhibitory effect on osteoclasts is mediated by a factor of low molecular mass (1-10,000 Da) present in the conditioned medium of the CRP 10/30 cells.[74] The identity of this factor remains to be determined. Similar inhibitory effects on resorption or osteoclast formation have since been reported using other osteoblast-like cells.[75,76]

MACROPHAGES

Like osteoclasts, macrophages in vitro are also particularly susceptible to the inhibitory effects of bisphosphonates in vitro. This includes inhibition of cell migration[77] and proliferation,[78-80] as well as effects that lead to inhibition of cell function or cytotoxicity.[77,78,81] Clodronate and pamidronate have also been shown to inhibit the ability of macrophages to resorb bone in vitro.[82,83] Since both macrophages and osteoclasts are highly pinocytic and phagocytic, the sensitivity of these cells to bisphosphonates most probably reflects their particular ability to internalize bisphosphonates intracellularly.[79] Indeed, the toxic and anti-proliferative effects of bisphosphonates on macrophages (but not nonphagocytic cells) can be enhanced either by encapsulation of bisphosphonate within liposomes, thus increasing the intracellular delivery of bisphosphonate,[80] or in the presence of bone mineral, which would also have the effect of concentrating the bisphosphonate.[82,83] We have recently shown that toxicity of bisphosphonates towards macrophages, like osteoclasts, is due to induction of apoptotic cell death.[84,85] The characteristic features of apoptosis (namely condensation of chromatin, nuclear condensation and fragmentation, together with activation of an endonuclease that catalyzes internucleosomal DNA fragmentation) can be identified in J774 and RAW264 mouse macrophage-like cells approximately 24 hours after treatment with 10 μM or higher of alendronate, pamidronate and ibandronate, or at concentrations of 1mM or higher of clodronate.

It is unclear whether bisphosphonates may also affect macrophages in vivo. Although clodronate, at least, does not appear to impair host defense mechanisms in humans,[86] some bisphosphonates (the aminohydroxyalkylbisphosphonates, such as alendronate and pamidronate) do cause a transient acute phase response following administration in humans, involving pyrexia, lymphopenia, decreased serum zinc and increased C-reactive protein.[87] These effects may be the result of a transient interaction of bisphosphonates with circulating monocytes and lymphocytes in the circulation or in the bone marrow, leading to the release of cytokines[88] such as interleukin-6, interleukin-1 and tumor necrosis factor, an effect that has been observed with human peripheral blood cells in vitro.[89] Paradoxically, these cytokines also stimulate osteoclast formation.[26] The release of such cytokines following exposure of monocytes and macrophages to bisphosphonates may also explain the increase in osteoclast number that has been observed in mice and rats following treatment with pamidronate, clodronate and etidronate.[3,90-94]

An alternative explanation is that some bisphosphonates may increase osteoclast formation by stimulating the synthesis of histamine in the bone marrow.[95]

Since bisphosphonates can clearly affect monocytes and macrophages, inhibition of bone resorption could partly occur by an indirect effect via these cells in the bone marrow, which are a source of resorption-stimulating cytokines.[79] An effect of bisphosphonates on the monocyte/macrophage system in vivo may also partly account for the anti-inflammatory effects of bisphosphonates that have been described in models of arthritis in animals[96-100] as well as in rheumatoid arthritis in humans,[101] since bisphosphonate released at sites of chronic inflammation and active bone resorption could affect bone-resident macrophages or synovial cells, thus leading to down-regulation of the inflammatory process.[97] In addition, since liposome-encapsulated clodronate is toxic to phagocytic macrophages but not other cells in vivo[102,103] this may lead to the use of liposomal bisphosphonates in the depletion of macrophages in inflammatory diseases such as rheumatoid arthritis.[104-106]

MODELS FOR IDENTIFYING THE MOLECULAR MECHANISMS OF ACTION OF BISPHOSPHONATES

There is an obvious need to clarify the exact molecular mechanisms by which bisphosphonates inhibit bone resorption. While osteoclasts or their precursors may be the major route through which bisphosphonates inhibit resorption, these cells are difficult to isolate in sufficient numbers or in pure cultures for biochemical and molecular studies. In vitro cellular models of osteoclasts may therefore hold the key to identifying the molecular targets for bisphosphonates. Cellular uptake of bisphosphonates by endocytosis is probably an important step in determining the susceptibility of cells to bisphosphonates. Whereas Felix et al have demonstrated cellular uptake of radiolabeled clodronate and etidronate in connective tissue cells,[107] we have used fluorescently-derivatized analogs of alendronate to visualize rapid and efficient endocytic uptake of bisphosphonate by osteoclasts and macrophages in vitro.[108] Macrophages therefore appear to be the most obvious model for osteoclasts, since both cell types are highly endocytic and capable of removing bone mineral. In addition to the numerous reports of toxic and anti-proliferative effects of bisphosphonates on macroph-ages[77-80,82,83] we have found that macrophage-like cells (such as the murine cell lines J774 and RAW264) also undergo apoptosis, like osteoclasts, following exposure to bisphosphonates.[84,85] We are be-ginning to identify some of the biochemical features of bis-phosphonate-induced apoptosis in these cells in an attempt to iden-tify the sequence of intracellular events that lead to apoptosis. Our preliminary results suggest that apoptosis may be the consequence of an effect on cellular metabolism, since apoptosis occurs follow-ing a lag period of about 24 hours after exposure to alendronate or pamidronate.[85] Meanwhile Mönkkönen et al have suggested that bisphosphonates may affect macrophages by depleting intracellular iron stores.[109]

We have also identified a more novel model system with which to identify the molecular basis for the effects of bisphosphonates, using amoebae of the cellular slime mold *Dictyostelium discoideum*. Bisphosphonates inhibit the growth of amoebae of this eukaryotic microorganism in a dose-dependent manner. Furthermore, the structure-activity relationships of bisphosphonates for inhibition of growth of the amoebae closely match those for inhibition of bone resorption.[110] Hence, clodronate and etidronate are amongst the least potent, with IC_{50}s for growth inhibition of 350 μM or more (Fig. 7.7). The aminoalkylbisphosphonates, including pamidronate, alendronate and neridronate, are much more potent (IC_{50}s in the range 30-200 μM), whereas the most potent anti-resorptive bisphosphonates such as risedronate and ibandronate inhibit *Dictyostelium* growth at concentrations of 10 μM or lower. Minor changes to the structure of bisphosphonates that have a dramatic effect on anti-resorptive potency also affect potency toward *Dictyostelium* in a similar manner. This includes alterations to the side chains[111] and to the phosphonate groups.[112] Hence, bis-phosphonates may interact with a specific target that is present in *Dictyostelium* and osteoclasts, and thus probably in all eukaryotic cells. Since phosphonophosphinate and bisphosphinate analogs of bisphosphonates are far less active both as growth inhibitors of *Dictyostelium* and as anti-resorptive agents, this interaction appears to require two phosphonate groups as well as side chains of a par-ticular conformation and chemical composition. Like macrophages

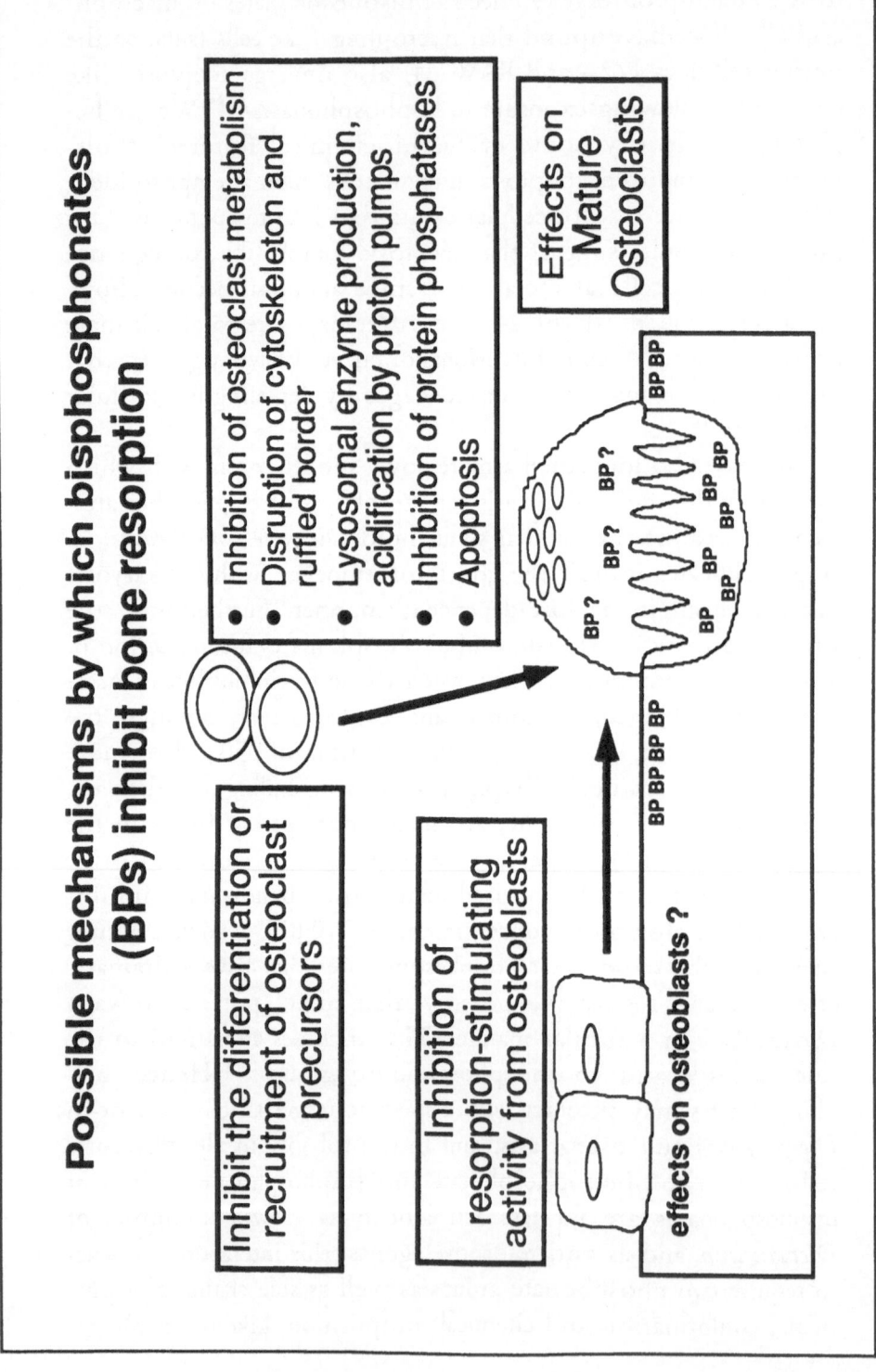

Fig. 7.6. Potential cellular mechanisms of action of bisphosphonates in bone.

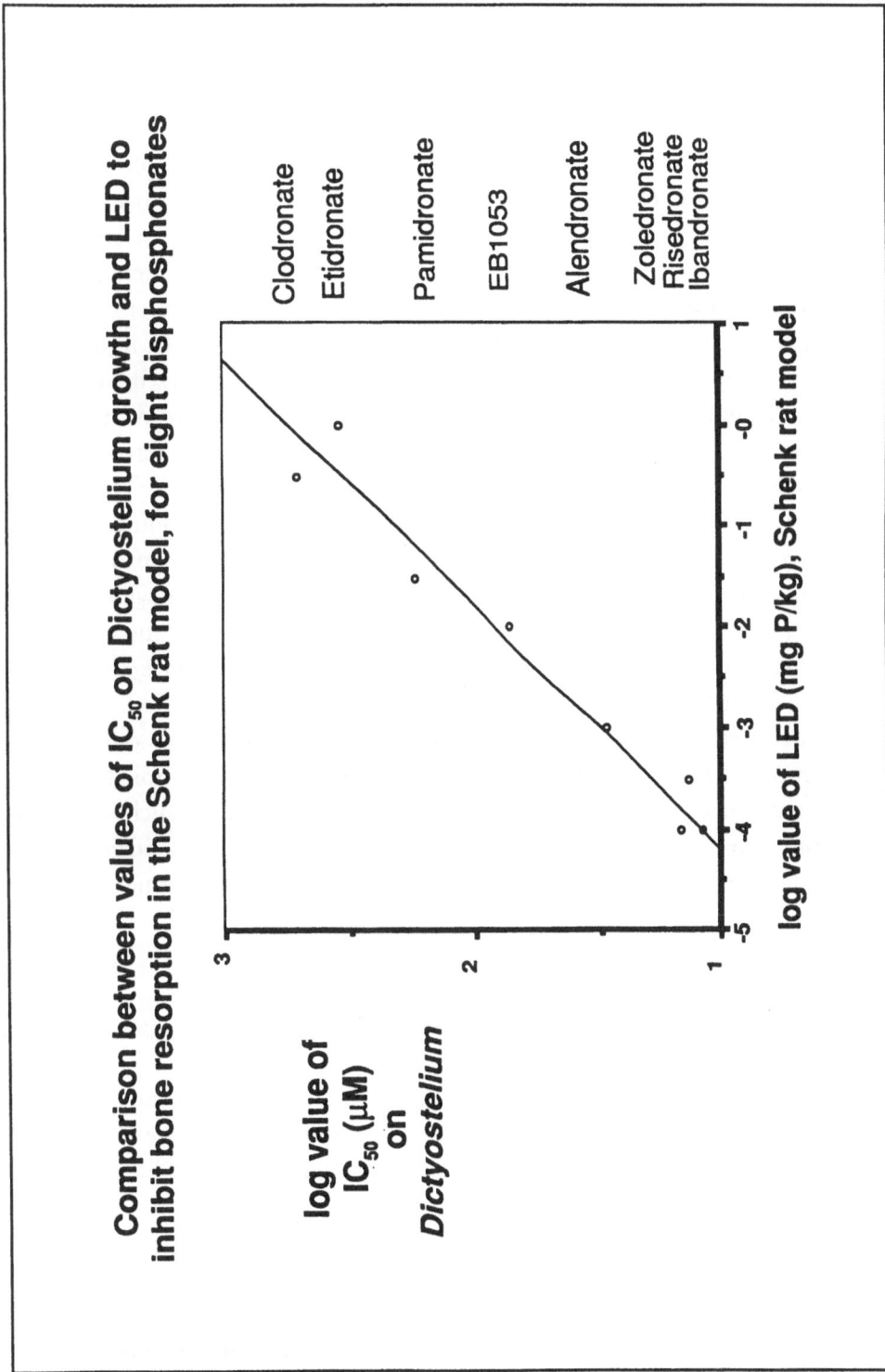

Fig. 7.7. Potency of bisphosphonates as inhibitors of growth in Dictyosteleum discoideum compared with inhibition of bone resorption in rats (Schenk model)[3].

and osteoclasts, the susceptibility of *Dictyostelium* to bisphosphonates is probably due to the ability of the cells to internalize bisphosphonates by endocytosis[108] since *Dictyostelium* amoebae that are prevented from carrying out endocytosis become markedly resistant to the growth-inhibitory effects of bisphosphonates.[113] The *Dictyostelium* model has already provided an insight into potential molecular mechanisms by which some bisphosphonates may act, since bisphosphonates of low potency (generally those with short side chains, such as clodronate, that are close analogs of PP_i) can be metabolized intracellularly into nonhydrolyzable analogs of ATP by a back-reaction catalyzed by certain aminoacyl-tRNA synthetases.[114,115] These observations demonstrated for the first time that at least some bisphosphonates are not necessarily metabolically inert, as has been suggested.[116] The formation of potentially toxic metabolites of some bisphosphonates could also account for the effects of these compounds on mammalian cells,[117] since the same aminoacyl-tRNA synthetase enzymes from cell-free extracts of human cells also appear capable of incorporating bisphosphonates into ATP analogs in vitro.[118] Since the more potent bisphosphonates such as pamidronate and risedronate are not metabolized, these bisphosphonates probably affect cells by a molecular mechanism which is different to that of clodronate and etidronate. The use of homogeneous cultures of cells, such as *Dictyostelium* and macrophages, will be of great benefit in the identification of biological molecules with which bisphosphonates interact.

CONCLUSIONS

Bisphosphonates may inhibit osteoclast-mediated bone resorption by a variety of mechanisms. This includes a direct effect on osteoclasts, due to pertubation of cellular metabolism or to induction of osteoclast cell death by apoptosis, or inhibition of the formation of osteoclasts from hemopoietic precursors. In addition, bisphosphonates may affect osteoclasts indirectly through the mediation of other bone cells such as osteoblasts. The molecular mechanisms by which these effects are brought about remain unclear, although bisphosphonates can affect a variety of proteins involved in cellular metabolism, signal transduction and cell function. Furthermore, it is likely that the mechanism of action of bisphosphonates that most resemble pyrophosphate (such as clodronate and etidronate) is different to that of the more potent anti-resorptive bisphosphonates. In vitro models of osteoclasts that reflect the biological properties of bisphosphonates, such as macrophages and *Dictyostelium* amoebae, will be instrumental in the final identification of the molecular mechanisms of action of these important drugs.

REFERENCES

1. Fleisch H, Russell RGG, Francis MD. Diphosphonates inhibit hydroxyapatite dissolution in vitro and bone resorption in tissue culture and in vivo. Science 1969; 165:1264-1266.
2. Reynolds JJ, Minkin C, Morgan DB et al. The effect of two diphosphonates on the resorption of mouse calvaria in vitro. Calcif Tiss Res 1972; 10:302-313.
3. Schenk R, Merz WA, Muhlbauer R et al. Effect of ethane-1-hydroxy-1,1-diphosphonate (EHDP) and dichloromethylene diphosphonate (Cl₂MDP) on the calcification and resorption of cartilage and bone in the tibial epiphysis and metaphysis of rats. Calc Tiss Res 1973; 11:196-214.
4. Smith R, Russell RGG, Bishop M. Diphosphonates and Paget's disease of bone. Lancet 1971; 1:945-947.
5. Fleisch H. Bisphosphonates, pharmacology and use in the treatment of tumour-induced hypercalcaemic and metastatic bone disease. Drugs 1991; 42:919-942.
6. Storm T, Thamsborg G, Steiniche T et al. Effect of intermittent cyclical etidronate therapy on bone mass and fracture rate in women with postmenopausal osteoporosis. New Engl J Med 1990; 322:1265-71.
7. Watts NB, Harris ST, Genant HK et al. Intermittent cyclical etidronate treatment of postmenopausal osteoporosis. N Engl J Med 1990; 323:73-79.
8. Fleisch H, Bisaz S. Isolation from urine of pyrophosphate, a calcification inhibitor. Am J Physiol 1962; 203:671-675.
9. Fleisch H, Neuman WF. Mechanisms of calcification: role of collagen, polyphosphates, and phosphatase. Am J Physiol 1961; 200:1296-1300.
10. Fleisch H, Russell RGG, Straumann F. Effect of pyrophosphate on hydroxyapatite and its implications in calcium homeostasis. Nature 1966; 212:901-903.
11. Francis MD. The inhibition of calcium hydroxyapatite crystal growth by polyphosphates. Calcif Tiss Res 1969; 3:151-162.

12. Fleisch H, Russell RGG, Francis MD. Diphosphonates inhibit formation of calcium phosphate crystals in vitro and pathological calcification in vivo. Science 1969; 165:1262-1264.
13. Fleisch HA, Russell RGG, Bisaz S et al. The inhibitory effect of phosphonates on the formation of calcium phosphate crystals in vitro and on aortic and kidney calcification in vivo. Eur J Clin Invest 1970; 1:12-18.
14. Russell RGG, Muhlbauer RC, Bisaz S et al. The influence of pyrophosphate, condensed phosphates, phosphonates and other phosphate compounds on the dissolution of hydroxyapatite in vitro and on bone resorption induced by parathyroid hormone in tissue culture and in thyroparathyroidectomised rats. Calc Tiss Res 1970; 6:183-196.
15. Jung A, Bisaz S, Fleisch H. The binding of pyrophosphate and two diphosphonates on hydroxyapatite crystals. Calc Tiss Res 1973; 11:269-280.
16. Ebrahimpour A, Francis MD. Bisphosphonate therapy in acute and chronic bone loss: physical chemical considerations in bisphosphonate-related therapies. In: Bijvoet O, Fleisch HA, Canfield RE, Russell RGG, eds. Bisphosphonate on bones. Elsevier Science B.V., 1995.
17. Shinoda H, Adamek G, Felix R et al. Structure-activity relationships of various bisphosphonates. Calcif Tiss Int 1983; 35:87-99.
18. Geddes AD, D'Souza SM, Ebetino FH et al. Bisphosphonates: structure-activity relationships and therapeutic implications. Bone and Mineral Res 1994; 8:265-306.
19. Schenk R, Eggli P, Fleisch H et al. Quantitative morphometric evaluation of the inhibitory activity of new aminobisphosphonates on bone resorption in the rat. Calc Tiss Int 1986; 38:342-349.
20. Muhlbauer RC, Bauss F, Schenk R et al. BM 21.0955, a potent new bisphosphonate to inhibit bone resorption. J Bone Min Res 1991; 6:1003-1011.
21. Boonekamp PM, Löwik CWGM, van der Wee-Pals LJA et al. Enhancement of the inhibitory action of APD on the transformation of osteoclast precursors into resorbing cells after dimethylation of the amino group. Bone and Mineral 1987; 2:29-42.
22. Van der Pluijm G, Binderup L, Bramm E et al. Disodium 1-hydroxy-3-(1-pyrrolidinyl)-propylidene-1,1-bisphosphonate (EB-1053) is a potent inhibitor of bone resorption in vitro and in vivo. J Bone Miner Res 1992; 7:981-986.
23. Sietsema WK, Ebetino FH, Salvagno AM et al. Antiresorptive dose-response relationships across three generations of bisphosphonates. Drugs Exptl Clin Res 1989; 15:389-396.
24. Green JR, Muller K, Jaeggi KA. Preclinical pharmacology of CGP

42'446, a new, potent, heterocyclic bisphosphonate compound. J Bone Miner Res 1994; 9:745-751.

25. Ebetino FH, Russell RGG. Metabolic bone disease: current therapies and future prospects with bisphosphonates and other agents. In: Sarel S, Mechoulam R, Agranat I, eds. Trends in Medicinal Chemistry '90. Oxford:Blackwell Scientific Publications, 1992: 293-298.

26. Mundy GR, Roodman GD. Osteoclast ontogeny and function. Bone and Mineral Res 1987; 5:209-279.

27. Zaidi M, Towhidul Alam ASM, Shankar VS et al. Cellular biology of bone resorption. Biol Rev 1993; 68:197-264.

28. Horton MA, Davies J. Perspectives: Adhesion receptors in bone. J Bone Miner Res 1989; 4:803-808.

29. Davies J, Warwick J, Totty N et al. The osteoclast functional antigen implicated in the regulation of bone resorption is biochemically related to the vitronectin receptor. J Cell Biol 1989; 109:1817-1826.

30. Lakkakorpi PT, Vaananen HK. Kinetics of the osteoclast cytoskeleton during the resorption cycle in vitro. J Bone Miner Res 1991; 6:817-826.

31. Baron R, Neff L, Louvard D et al. Cell-mediated extracellular acidification and bone resorption: evidence for a low pH in resorbing lacunae and localization of a 100-kD lysosomal membrane protein at the osteoclast ruffled border. J Cell Biol 1985; 101:2210-2222.

32. Blair HC, Teitelbaum SL, Ghiselli R et al. Osteoclastic bone resorption by a polarized vacuolar proton pump. Science 1989; 245:855-857.

33. Väänänen HK, Karhukorpi E-K, Sundquist K et al. Evidence for the presence of a proton pump of the vacuolar H^+-ATPase type in the ruffled borders of osteoclasts. J Cell Biol 1990; 111:1305-1311.

34. Baron R. Molecular mechanisms of bone resorption by the osteoclast. Anat Rec 1989; 224:317-324.

35. Bisaz S, Jung A, Fleisch H. Uptake by bone of pyrophosphate, diphosphonates and their technetium derivatives. Clin Sci Mol Med 1978; 54:265-272.

36. Sato M, Grasser W, Endo N et al. Bisphosphonate action. Alendronate localisation in rat bone and effects on osteoclast ultrastructure. J Clin Invest 1991; 88:2095-2105.

37. Flanagan AM, Chambers TJ. Inhibition of bone resorption by bisphosphonates: interactions between bisphosphonates, osteoclasts and bone. Calcif Tiss Int 1991; 49:407-415.

38. Rowe DJ, Hausmann E. The alteration in osteoclast morphology by diphosphonates in bone organ culture. Calc Tiss Res 1976; 20:53-60.

39. Rowe DJ, Hausmann E. The effects of calcitonin and colchicine on the cellular response to diphosphonate. Br J Exp Path 1980; 61:303-309.

40. Rowe DJ, Hays SJ. Inhibition of bone resorption by difluoromethylene diphosphonate in organ culture. Metab Bone Dis and Rel Res 1983; 5:13-16.

41. Flanagan AM, Chambers TJ. Dichloromethylenebisphosphonate (Cl$_2$MBP) inhibits bone resorption through injury to osteoclasts that resorb Cl$_2$MBP-coated bone. Bone and Mineral 1989; 6:33-43.

42. Wyllie AH, Kerr JFR, Currie AR. Cell death: the significance of apoptosis. Int Rev Cytol 1980; 68:251-306.

43. Bortner CD, Oldenburg NBE, Cidlowski, JA. The role of DNA fragmentation in apoptosis. Trends Cell Biol 1995; 5:21-26.

44. Hughes DE, Wright KR, Uy HL et al. Bisphosphonates promote apoptosis in murine osteoclasts in vitro and in vivo. J Bone Miner Res 1995; 10:1478-1487.

45. Selander KS, Härkönen PL, Mönkkönen J et al. The mode of death induced by clodronate in isolated osteoclasts is apoptosis. Calc Tiss Int 1995; 56:450 (abstract).

46. Hughes DE, Wiltschke K, Mundy GR et al. Estrogen promotes osteoclast apoptosis in vitro and in vivo. Bone 1995; 16 (S1):93S (abstract).

47. Hughes DE, Wright KR, Mundy GR et al. TGFβ1 induces osteoclast apoptosis in vitro. J Bone Miner Res 1994; 9(Suppl 1):S138 (abstract).

48. Carano A, Teitelbaum SA, Konsek JD et al. Bisphosphonates directly inhibit the bone resorption activity of isolated avian osteoclasts in vitro. J Clin Invest 1990; 85:456-461.

49. Felix R, Fleisch H. Increase in alkaline phosphatase activity in calvaria cells cultured with diphosphonates. Biochem J 1979; 183:73-81.

50. Morgan DB, Monod A, Russell RGG et al. Influence of dichloromethylene diphosphonate and calcitonin on bone resorption, lactate production and phosphatase and pyrophosphatase content of mouse calvaria treated with parathyroid hormone in vitro. Calc Tiss Res 1973; 13:287-294.

51. Fast DE, Felix R, Dowse C et al. The effects of diphosphonates on the growth and glycolysis of connective-tissue cells in culture. Biochem J 1978; 172:97-107.

52. Ende JJ, van Rooijen HJM. Some effects of EHDP and Cl$_2$MDP on the metabolism of mouse calvaria in tissue culture. Proc Kon Ned Akad Wet 1979; C82:43-54.

53. Amin D, Cornell SA, Gustafson SK et al. Bisphosphonates used for the treatment of bone disorders inhibit squalene synthase and

cholesterol biosynthesis. J Lipid Res 1992; 33:1657-1663.

54. Felix R. Studies with isolated cells and systems. In: Bijvoet O, Fleisch HA, Canfield RE, Russell RGG, eds. Bisphosphonate on Bones. Elsevier Science B.V. 1995:189-204.

55. Miller SC, Jee WSS. The effect of dichloromethylene diphosphonate, a pyrophosphate analog, on bone and bone cell structure in the growing rat. Anatomical Rec 1979; 193:439-462.

56. Plasmans CMT, Jap PHK, Kuijpers W et al. Influence of a diphosphonate on the cellular aspect of young bone tissue. Calcif Tissue Int 1980; 32:247-256.

57. Sato M, Grasser W. Effects of bisphosphonates on isolated rat osteoclasts as examined by reflected light microscopy. J Bone Miner Res 1990; 5:31-40.

58. Murakami H, Takahashi N, Sasaki T et al. A possible mechanism of the specific action of bisphosphonates on osteoclasts: tiludronate preferentially affects polarized osteoclasts having ruffled borders. Bone 1995; 17:137-144.

59. Boyce BF, Yoneda T, Lowe C et al. Requirement of pp60src expression for osteoclasts to form ruffled borders and resorb bone in mice. J Clin Invest 1992; 90:1622-1627.

60. Schmidt A, Opas E, Rutledge SJ et al. Alendronate inhibition of protein tyrosine phosphatase activity. Bone 1995; 17:604 (abstract).

61. Murakami H, Takahashi N, Udagawa N et al. Tiludronate inhibits protein tyrosine phosphatase activity in osteoclasts. Bone 1995; 16 (S1):114S (abstract).

62. David P, Nguyen H, Barbier A et al. Tiludronate is a potent and specific inhibitor of the osteoclast vacuolar H^+ ATPase. Bone 1995; 16 (S1):166S (abstract).

63. Zimolo Z, Wesolowski G, Rodan GA. Acid extrusion is induced by osteoclast attachment to bone. J Clin Invest 1995; 96:2277-2283.

64. Felix R, Russell RGG, Fleisch H. The effect of several diphosphonates on acid phosphohydrolases and other lysosomal enzymes. Biochim Biophys Acta 1976; 429:429-438.

65. Lerner UH, Larsson A. Effects of four bisphosphonates on bone resorption, lysosomal enzyme release, protein synthesis and mitotic activities in mouse calvarial bones in vitro. Bone 1987; 8:179-189.

66. Löwik C, van der Pluijm G. Mechanisms of action of bisphosphonates: studies with bone culture systems. In: Bijvoet O, Fleisch HA, Canfield RE, Russell RGG, eds. Bisphosphonate on Bones. Elsevier Science B.V. 1995:155-170.

67. Hughes DE, MacDonald BR, Russell RGG et al. Inhibition of osteoclast-like cell formation by bisphosphonates in long-term cultures of human bone marrow. J Clin Invest 1989; 83:1930-1935.

68. Boonekamp PM, van der Wee-Pals LJA, van Wijk-van Lennep MLL

et al. Two modes of action of bisphosphonates on osteoclastic re-sorption of mineralized matrix. Bone and Mineral 1986; 1:27-39.

69. Löwik CWGM, van der Pluijm G, van der Wee-Pals LJA et al. Migration and phenotypic transformation of osteoclast precursors into mature osteoclasts: the effect of a bisphosphonate. J Bone Miner Res 1988; 3:185-192.

70. Papapoulos SE, Hoekman K, Lowik CWGM et al. Application of an in vitro model and a clinical protocol in the assesment of the potency of a new bisphosphonate. J Bone Miner Res 1988; 4:775-781.

71. Guenther HL, Guenther HE, Fleisch H. Effects of 1-hydroxyethane-1,1-diphosphonate and dichloromethanediphosphonate on rabbit articular chondrocytes in culture. Biochem J 1979; 184:203-214.

72. D'Souza SM, Orcutt CM, Ibbotson KJ. Inhibitory effects of bisphosphonates on rat osteoblast-like cells: correlation with anti-resorptive potency. J Bone Miner Res 1990; 5:S90 (abstract).

73. Sahni M, Guenther H, Fleisch H et al. Bisphosphonates act on rat bone resorption through the mediation of osteoblasts. J Clin Invest 1993; 91:2004-2011.

74. Vitte C, Fleisch H, Guenther HL. Osteoblasts mediate the bisphosphonate inhibition on bone resorption through synthesis of an inhibitor of osteoclastic resorption. Bone 1995; 17:602 (abstract).

75. Yu X, Scholler J, Foged NT. Pretreatment of an osteoblast-like cell line with bisphosphonates inhibits its PTH-stimulated induction of osteoclastic bone resorption. J Bone Miner Res 1994; 9(Suppl 1):S232 (abstract).

76. Niskikawa M, Akatsu T, Katayama Y et al. Bisphosphonates act on osteoblastic cells and inhibit osteoclastic cell formation in mouse marrow culture. Bone 1995; 16 (S1):167S (abstract).

77. Stevenson PH, Stevenson JR. Cytotoxic and migration inhibitory effects of bisphosphonates on macrophages. Calcif Tiss Int 1986; 38:227-233.

78. Cecchini MG, Felix R, Fleisch H et al. Effect of bisphosphonates on proliferation and viability of mouse bone marrow-derived mac-rophages. J Bone Miner Res 1987; 2:135-142.

79. Cecchini MG, Fleisch H. Bisphosphonates in vitro specifically in-hibit, among the hematopoietic series, the development of the mouse mononuclear phagocyte lineage. J Bone Miner Res 1990; 5: 1019-1027.

80. Mönkkönen J, Taskinen M, Auriola SOK et al. Growth inhibition of macrophage-like and other cell types by liposome-encapsulated, calcium-bound and free bisphosphonates in vitro. J Drug Targetting 1994; 2:299-308.

81. De Vries E, van der Weij JP, v.d. Veen CJP et al. In vitro effect

of (3-amino-1-hydroxypropylidene)-1,1-bisphosphonic acid (APD) on the function of mononuclear phagocytes in lymphocyte proliferation. Immunology 1982; 47:157-163.

82. Chambers TJ. Diphosphonates inhibit bone resorption by macrophages in vitro. J Pathol 1980; 132:255-262.

83. Reitsma PH, Teitelbaum SL, Bijvoet OLM et al. Differential action of the bisphosphonates (3-amino-1-hydroxypropylidene)-1,1-bisphosphonate (APD) and disodium dichloromethylidene bisphosphonate (Cl_2MDP) on rat macrophage-mediated bone resorption in vitro. J Clin Invest 1982; 70:927-933.

84. Rogers MJ, Chilton KM, Coxon FP et al. Biosphosphonates induce apoptosis in mouse macrophage-like cells by a nitric oxide independent mechanism. J Bone Miner Res 1996; in press.

85. Coxon FP, Russell RGG, Rogers MJ. Pathways of bisphosphonate-induced apoptosis in murine macrophage-like cells. Bone 1995; 17:600 (abstract).

86. Markkula R, Repo, H, Leirisal M et al. Effect of dichloromethylene diphosphonate (Cl_2MDP) on immune function in breast cancer patients with bone metastases. Cancer Immunol Immunother 1983; 15:159-161.

87. Adami S, Bhalla AK, Dorizzi R et al. The acute-phase response after bisphosphonate administration. Calcif Tiss Int 1987; 41: 326-331.

88. Schweitzer DH, Oostendorp-van de Ruit M, van der Pluijm G et al. Interleukin-6 and the acute phase response during treatment of patients with Paget's disease with the nitrogen-containing bisphosphonate dimethylaminohydroxypropylidene bisphosphonate. J Bone Miner Res 1995; 10:956-962.

89. Pioli G, Hughes DE, Mian M et al. Bisphosphonates stimulate the production of cytokines by human peripheral blood mononuclear cells in vitro. Calcif Tiss Int 1990; 46(Suppl. 2):A54 (abstract).

90. Miller SC, Jee WSS, Kimmel DB et al. Ethane-1-hydroxy-1,1-diphosphonate (EHDP) effects on incorporation and accumulation of osteoclast nuclei. Calc Tissue Res 1977; 22:243-252.

91. Evans RA, Howlett CR, Dunstan CR et al. The effect of long-term low-dose diphosphonate treatment on rat bone. Clin Orthop Rel Res 1982; 165:290-299.

92. Marshall MJ, Holt I, Davie MWJ. Osteoclast recruitment in mice is stimulated by (3-amino-1-hydroxypropylidene)-1,1-bisphosphonate. Calcif Tissue Int 1990; 52:21-25.

93. Marshall MJ, Wilson AS, Davie MWJ. Effects of (3-amino-1-hydroxypropylidene)-1,1-bisphosphonate on mouse osteoclasts. J Bone Miner Res 1993; 5:955-962.

94. Katoh Y, Tsuji H, Matsui K et al. Effects of ethane-1-hydroxy-

1,1-diphosphonate on cell differentiation and proteoglycan and calcium metabolism in the proximal tibia of young rats. Bone 1991; 12:59-65.

95. Endo Y, Nakamura M, Kikuchi T et al. Aminoalkylbisphosphonates, potent inhibitors of bone resorption, induce a prolonged stimulation of histamine synthesis and increase macrophages, granulocytes and osteoclasts in vivo. Calcif Tissue Int 1993; 52:248-254.

96. Francis MD, Flora L, King WR. The effects of disodium ethane-1-hydroxy-1,1-diphosphonate on adjuvant induced arthritis in rats. Calc Tiss Res 1972; 9:109-121.

97. Francis MD, Horancik K, Boyce RW. NE 58095 A diphosphonate which prevents bone erosion and preserves joint architecture in experimental arthritis. Int J Tissue React 1989; 11:239-252.

98. Flora L. Comparative antiinflammatory and bone protective effects of two diphosphonates in adjuvant arthritis. Arthritis Rheum 1979; 22:340-346.

99. Barbier A, Breliere JC, Remandet B et al. Studies on the chronic phase of adjuvant arthritis: effect of SR 41319, a new diphosphonate. Ann Rheum Dis 1986; 45:67-74.

100. Dunn CJ, Galinet LA, Wu H et al. Demonstration of novel anti-arthritic and anti-inflammatory effects of diphosphonates. J Pharmacol Exp Ther 1993; 266:1691-1698.

101. Bijvoet OLM, Frijlink WB, Jie K et al. APD in Paget's disease of bone. Role of the mononuclear system? Arthritis Rheum 1980; 23:1193-1204.

102. Van Rooijen N. The liposome mediated macrophage 'suicide' technique. J Immunol Methods 1989; 124:1-6.

103. Van Rooijen N, Kors N, van den Ende M et al. Depletion and repopulation of macrophages in spleen and liver of rat after intravenous treatment with liposome-encapsulated dichloromethylene diphosphonate. Cell Tissue Res 1990; 260:215-222.

104. Huitinga I, van Rooijen N, de Groot CJA et al. Suppression of experimental allergic encephalomyelitis in Lewis rats after elimination of macrophages. J Exp Med 1990; 172:1025-1033.

105. Van Lent PLEM, van den Bersselaar L, van den Hoek AEM et al. Reversible depletion of synovial lining cells after intraarticular treatment with liposome encapsulated dichloromethylene diphosphonate. Rheumatol Int 1993; 13:21-30.

106. Kinne RW, Schmidt-Weber CB, Hoppe R et al. Long-term amelioration of rat adjuvant arthritis following systemic elimination of macrophages by clodronate-containing liposomes. Arthritis Rheum 1995; 38:1777-1790.

107. Felix R, Guenther HL, Fleisch H. The subcellular distribution of [^{14}C]dichloromethylenebisphosphonate and [^{14}C]1-hydroxyethyli-

dene-1,1-bisphosphonate in cultured calvaria cells. Calcif Tiss Int 1984; 36:108-113.

108. Chestnut MH, Rogers MJ, Watts DJ et al. Cellular uptake of bisphosphonates: localisation using fluorescently-labelled alendronate. Bone 1995; 17:599 (abstract).

109. Mönkkönen J, Heath TD. The effects of liposome-encapsulated and free clodronate on the growth of macrophage-like cells in vitro: the role of calcium and iron. Calcif Tissue Int 1993; 53:139-146.

110. Rogers MJ, Watts DJ, Russell RGG et al. Inhibitory effects of bisphosphonates on growth of amoebae of the cellular slime mould Dictyostelium discoideum. J Bone Miner Res 1994; 9:1029-1039.

111. Rogers MJ, Xiong X, Brown RJ et al. Structure-activity relationships of new heterocycle-containing bisphosphonates as inhibitors of bone resorption and as inhibitors of growth of Dictyostelium discoideum amoebae. Mol Pharm 1995; 47:398-402.

112. Rogers MJ, Xiong X, Brown RJ et al. Alterations to the phosphonate groups of bisphosphonates decrease their potency as inhibitors of bone resorption and as inhibitors of Dictyostelium growth. Bone and Mineral 1994; 25(Suppl. 1):S69 (abstract).

113. Xiong X, Rogers MJ, Ji X, Russell RGG et al. Inhibition of growth of Dictyostelium amoebae is dependent on the cellular uptake of bisphosphonates by pinocytosis. Bone 1994; 15:246 (abstract).

114. Rogers M, Russell RGG, Blackburn GM et al. Metabolism of halogenated bisphosphonates by the cellular slime mould Dictyostelium discoideum. Biochem Biophys Res Commun 1992; 189:414-423.

115. Rogers MJ, Ji X, Russell RGG et al. Incorporation of bisphosphonates into adenine nucleotides by amoebae of the cellular slime mould Dictyostelium discoideum. Biochem Journal 1994; 303:303-311.

116. Michael WR, King WR, Wakim, JM. Metabolism of disodium ethane-1-hydroxy-1,1-diphosphonate (disodium etidronate) in the rat, rabbit, dog and monkey. Toxicol Appl Pharmacol 1972; 21:503-515.

117. Frith JC, Russell RGG, Blackburn GM et al. The anti-resorptive drug clodronate is metabolised to a non-hydrolysable ATP analogue by mammalian cells in vitro. Biochem Soc Trans 1996; 24; in press (abstract).

118. Rogers MJ, Brown RJ, Modkin V et al. Bisphosphoonates are incorporated into adenine nucleotides by human aminoacyl-tRNA synthetase enzymes. Biochem Biophys Res Commun 1996; in press.

CHAPTER 8

FAMILIAL EXPANSILE OSTEOLYSIS: A GENETIC MODEL OF PAGET'S DISEASE

Anne E. Hughes and R. John Barr

INTRODUCTION

Familial expansile osteolysis (FEO) is a severe hereditary bone dysplasia with some similarities to Paget's disease of bone.[1] It is inherited as an autosomal dominant trait through a large five-generation kindred in Northern Ireland (Fig. 8.1). There are at present 37 living members of this family who are known to have inherited the FEO gene, and a further 21 young members of generation V who are untested and who carry a 50% risk.

Affected individuals show both focal and generalized skeletal changes. The focal lesions develop in early adulthood and are usually located in the long bones. These lesions show increased osteoblast and osteoclast activity, leading to progressive medullary and cortical expansion of the bone. In the later stages the whole bone is expanded with few remaining histological similarities to the original tissue. Radiologically, there is a generalized alteration in the trabecular pattern of bone.[2] The focal lesions often cause painful and disabling deformity and have a tendency to pathological fracture. There is some variation in the severity of FEO

The Molecular Biology of Paget's Disease, edited by Paul T. Sharpe.
© 1996 R.G. Landes Company.

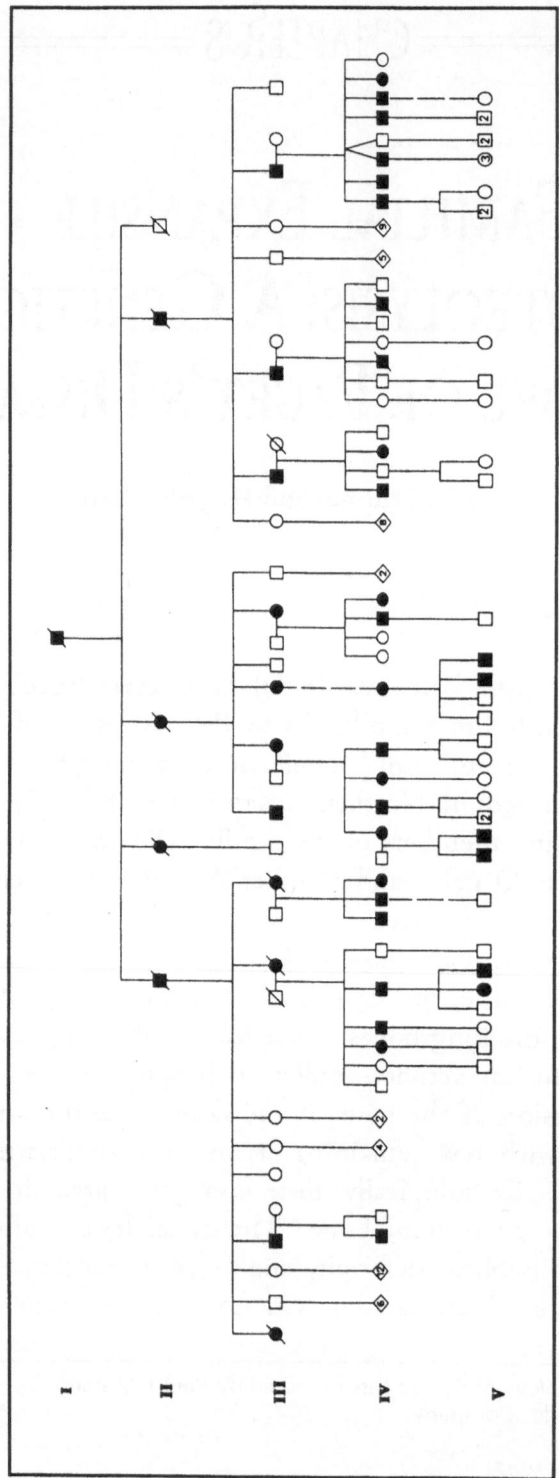

Fig. 8.1. Pedigree of the Northern Irish family with familial expansile osteolysis.

but amputation of affected limbs is often necessary. There is no satisfactory treatment for this disease.

The bone lesions show histological similarity to those in Paget's disease of bone, although their distribution and age of onset are different. It is possible that FEO is an autosomally inherited disorder at one end of the spectrum of Paget's disease of bone.[3] Paget's disease has a genetic component[4] but is difficult to study because of the late age of onset of focal bone lesions. It is hoped that identification of the FEO gene may assist the understanding of the etiology of Paget's disease of bone.

FAMILIAL EXPANSILE OSTEOLYSIS

CLINICAL ASPECTS

The skeletal changes of FEO are most commonly seen in the peripheral skeleton and only rarely in the limb girdles or axial skeleton. Lower limbs are more commonly affected than upper limbs, the tibia being the most commonly affected bone. The site of onset within each bone is variable as is the age of onset, which may be as early as 15 years. In a few individuals the disease is unifocal whereas in the majority it is multifocal. Lesions may also progress at differing rates. In those who are severely affected, the increasing lysis and expansion in the bone causes progressive deformity which is compounded by the deformity associated with secondary fracture. Although the bone is abnormal, fracture healing is not prolonged. In an unfortunate few, malignant transformation of longstanding lesions has occurred. This is an inevitably fatal complication and has accounted for three deaths in as many recent years. The rate at which the disease front advances along the shaft of a bone is unaffected by surgery or trauma and varies from 6.5-22 mm per year (mean 13.3 mm). It is of interest to note that the advancing lytic front in Paget's disease averages 7 mm per year.

The skeletal problems are frequently preceded by conductive type deafness which is present in more than 95% of affected patients. Deafness may occur in the first decade, but its onset is more common in the teenage years. Loosening and fracture of the teeth follows once the secondary dentition has arrived. With time the deafness pattern changes to a mixed picture and premature total tooth loss is the rule. With few exceptions, those affected will develop all features of the disease although the extent and severity of each may vary from one person to another and even within an individual.

PATHOLOGY

Familial expansile osteolysis is characterized by multiple focal lytic areas, altered modeling of some bones and a trabecular pattern alteration seen throughout the skeleton.

GENERALIZED ABNORMALITY

Disordered modeling is commonly seen in the long bones, particularly the humerus, radius, ulna and tibia. There are subtle alterations in shape and an alteration in the trabecular pattern in the form of a coarse "fish-net" character most commonly seen in metaphyseal bone. These changes are seen in all affected patients.

FOCAL ABNORMALITY

Radiologically, FEO is shown to be an exclusively lytic process. Three grades of focal change are recognized[2] (Fig. 8.2). Grade I change represents a lytic area in bone which is greater than 5 millimeters in diameter and which may cause cortical erosion but not cortical expansion. In grade II there is progression with subsequent expansion of the affected portion of bone and loss of the "normal" trabecular pattern. In grade III there is associated bony deformity due to mechanical failure of the affected bone. Another finding of note is that of an abnormal radioisotope bone scan. Using Technetium 99 labeled methylene diphosphonate, a proportionately greater bilateral uptake is observed in the proximal tibia when compared to the distal femur. This feature does not occur in unaffected members of the family nor has it been observed in any other bone disease. There is usually an increase in uptake at the sites of focal radiographic abnormalities.

BONE HISTOPATHOLOGY

Early stage biopsies show similarity to Paget's disease,[5] however, later stages are less compatible with this diagnosis. As the disease progresses the matrix becomes more scanty, with increased

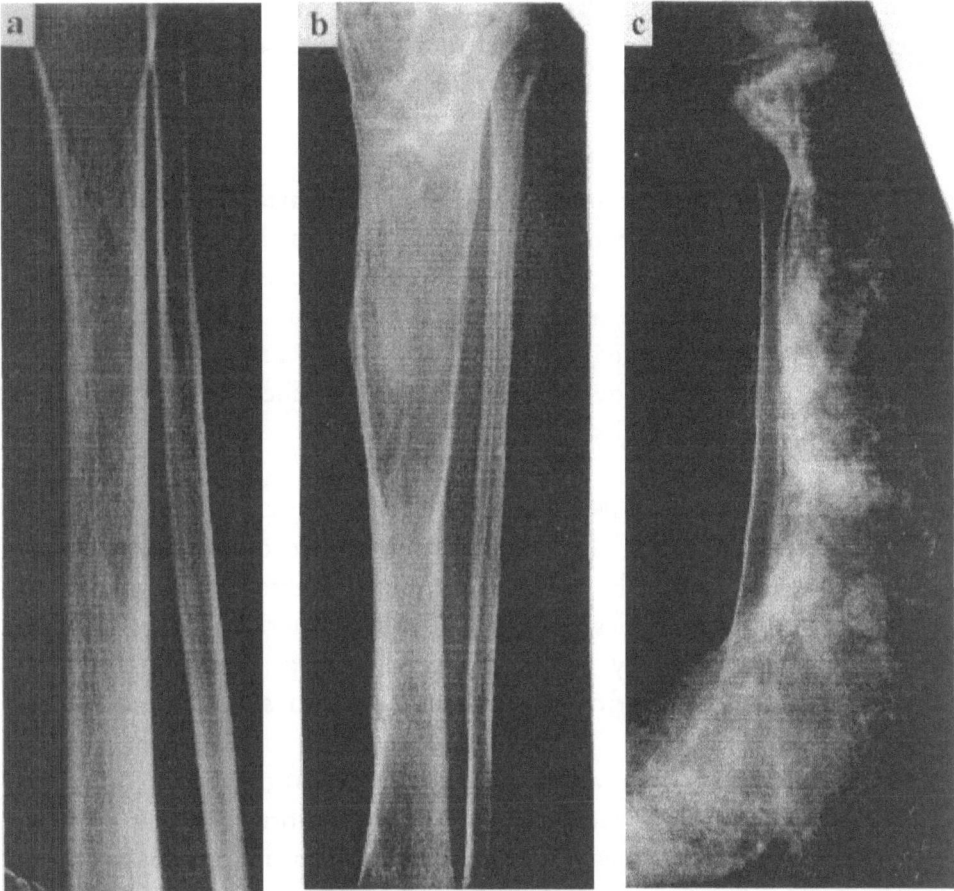

Fig. 8.2 a-c. Focal radiographic grades I-III in an FEO lesion (from left to right).

but ineffective osteoblast activity in the advanced stages. The bone trabeculae become disorganized with a corresponding increase in the fibrous tissue and more extensive vascularity. End stage disease is represented by almost complete fatty replacement of medullary bone with almost total loss of cortex, leaving few remaining features of the original bony structure.

It has also been noted that multiple osteoclasts (as in Paget's disease) are found adjacent to the matrix in diseased bone. When viewed under the electron microscope these osteoclasts are observed to have nuclear but no cytoplasmic viral-like microcylindrical inclusion bodies.[6] This is in contrast to Paget's disease in which both nuclear and cytoplasmic viral-like inclusions have been observed.

BIOCHEMISTRY

The serum alkaline phosphatase (ALP) is variably elevated in most affected patients with a logarithmic relationship between the serum ALP and the percentage of skeletal involvement. The urinary hydroxyproline is also elevated. These tests indicate an increased activity of osteoblasts and osteoclasts, respectively, thus indicating a high turnover bone disease. No other hematological or hormonal abnormality has been noted.

PATHOLOGY OF HEARING LOSS

Hearing loss is initially conductive in nature due to a stiffening of the ossicles in the middle ear. With progression of the disease the picture becomes one of a mixed conductive/sensory neural type as assessed by full audiometric testing. Tympanograms which assess the integrity of the ossicular chain suggest discontinuity.

At exploratory operation in 10 patients, abnormalities were observed in the long process of the incus, the most vulnerable part of the ossicular chain, with either absence, thinning or fibrous replacement. Thirty percent of patients had fixation of the stapedial footplate at the oval window and all showed sparing of the malleus. The sensory neural deafness may be caused either by the stapedial fixation as described or by compression of the auditory nerve as it leaves the inner ear through the auditory canal. In contrast to Paget's disease otosclerosis is not a feature of this disease.

DENTAL PATHOLOGY

Progressive tooth mobility precedes spontaneous tooth fracture with resultant premature loss of dentition. In general the tooth degeneration is painless unless infection or pulp necrosis supervenes. The cause of this degeneration with associated increased mobility is the marked external resorption at the apical and cervical regions of the teeth. Cervical erosion is most common and present in nearly all affected teeth. After decoronation the resorptive process continues and destroys the root thus making it unnecessary to perform root extraction. There is no other disease process known that causes apical and cervical resorption involving multiple teeth.[7]

DIAGNOSIS OF FEO

Until linkage with the FEO gene was established[8] a firm diagnosis of FEO was difficult to make before the clinical picture and associated investigations had become markedly abnormal. In a few cases diagnosis was possible in the teenage years, but in the majority it was not possible until the third decade. By this stage many affected members had married and had children of their own, each of whom had a 50% risk of inheriting the FEO gene.

Major and minor diagnostic criteria have been identified (Table 8.1). The presence of three criteria, at least one being major, were considered diagnostic. Minor features on their own are considered insufficient to establish diagnosis as there is a possibility that they may be present in conditions other than FEO. Now that linkage with the FEO gene has been established diagnosis can be made on the analysis of a blood sample by the polymerase chain

Table 8.1. Major and minor diagnostic criteria for familial expansile osteolysis

	Minor	Major
Otological	hearing loss with a compliant tympanogram	operative finding of fibrous replacement of the long process of the incus
Radiographic	generalized changes	focal changes I-III
Radioisotope		increased uptake below the knee
Biochemical	elevated serum ALP and urinary HPr	
Histology		biopsy confirming histological disease
Dental	cervical and apical resorption, single tooth involvement	cervical and apical resorption, multiple tooth involvement

reaction in association with known genetic markers. Where previously diagnosis was not possible until the disorder was clearly established, presymptomatic and prenatal diagnoses are now possible.

TREATMENT OF FEO

The majority of family members live within the Belfast area which has coordinated hospital and community care. A special clinic is held on a regular basis to deal with the special needs of FEO patients. This clinic is attended by a consultant orthopedic surgeon, a medical secretary, a nursing sister and a health visitor, so that continuity of care between the hospital and the community is ensured. The close cooperation with the social and community nursing service is essential.[9] Until such times as molecular methods are established to control or preferably to eliminate FEO, specific therapy with the aim of arresting or modifying the disease process is provided. Only by addressing the underlying cellular imbalance can we hope to delay or arrest the advancing disease front, cause regression of existing lesions and prevent the development of new focal lesions. Treatments such as analgesics, orthoses and surgery are also directed toward the relief of symptoms or complications.

All current and past specific treatments are based on the premise that there is histological similarity to Paget's disease. Calcitonin and dichloro-methylene-diphosphonate have been administered.[10,11] Although each showed an initial biochemical response, this was not sustained and in the long term failed to control either the symptoms or the progression of the disease. Radiotherapy to focal lesions in one patient met with little success, giving less than a week's relief of symptoms. More recently 10 patients with FEO were treated with intravenous 3-amino-1-hydroxypropylidine-1, 1-diphosphonate (APD). There was a significant fall in ALP levels up to and including one year following treatment with APD in all patients. A similar trend was observed in the urinary hydroxyproline:creatinine ratio. There was reduced scintographic activity in the focal lesions of six patients and radiological improvement in four (Fig. 8.3), each of whom was also symptomatically improved. No patient reported a sustained deterioration in symptoms.[12] To date APD has been shown to be the most effective

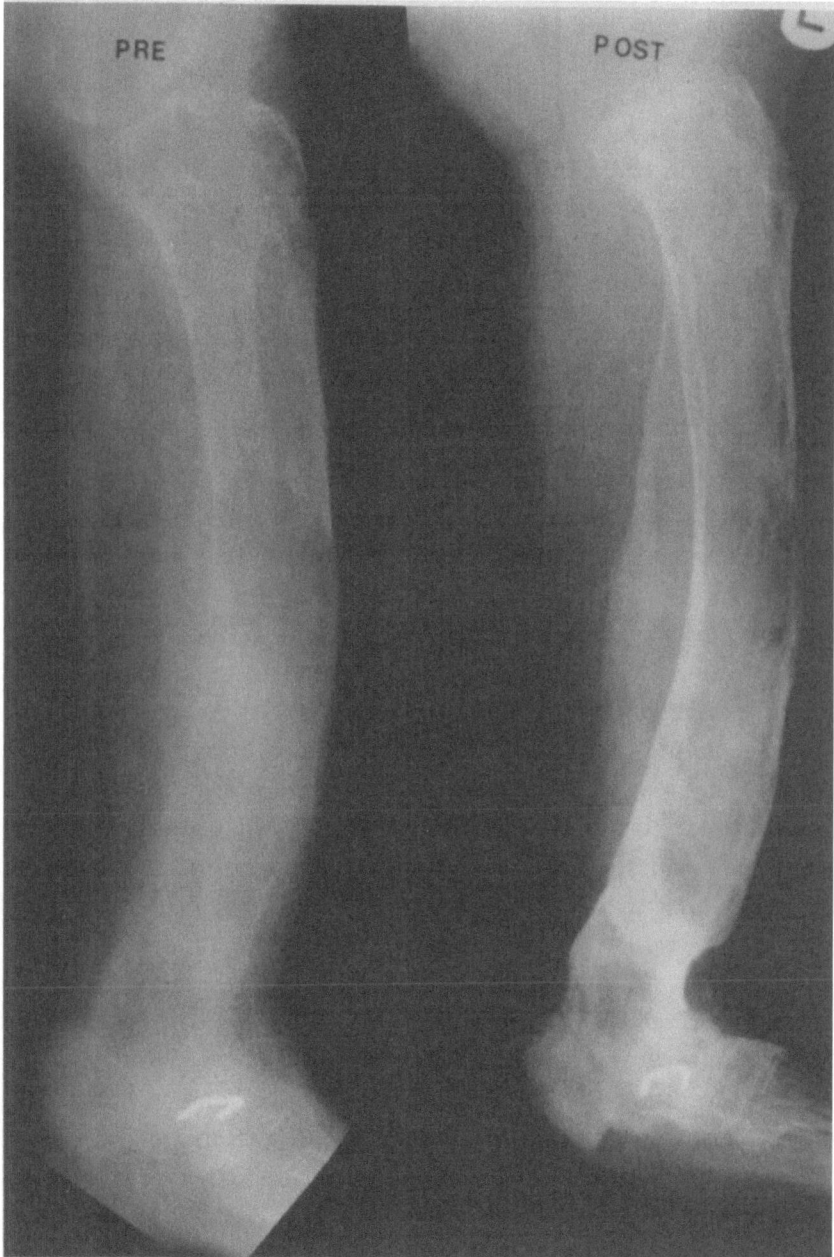

Fig. 8.3. Lateral pre- and post-treatment radiographs of the tibia of an FEO patient. Note the general increase in density and the better definition of cortical bone posteriorly in the post-treatment radiograph.

means of maintaining biochemical normality and the possibility of "booster" doses is being considered. It is hypothesized that with this strategy it may be possible to arrest or reverse the progress of focal lesions and thus reduce the long term morbidity associated with FEO. In addition patients who are identified as carrying the FEO gene may be offered APD as prophylaxis. Inevitably new treatments will become available for Paget's disease[13] and these will be assessed in due course.

GENETICS OF FEO

INHERITANCE OF FEO

FEO is inherited as an autosomal dominant trait through the kindred shown in Figure 8.1. The progenitor was born in Ireland in 1865 and had an arm amputated for a bony disorder before dying as a result of osteosarcoma. His sister had a leg amputated, possibly for the same disorder. A total of 46 individuals in five generations are now known to have inherited FEO, and many additional children in generation V are at risk of developing the disorder in adulthood. The FEO gene shows complete penetrance and no children of unaffected parents have symptoms. Diagnosis used to be difficult before the third decade because of the variable expression and age of onset of focal bone lesions. This problem has now been resolved by genetic advances. There is a considerable excess of males in the family (male:female ratio 1.7:1). Sixty percent of at risk members of either sex have inherited FEO. This is not, however, significantly higher statistically than the expected 50% for an autosomal dominant trait.

Two additional smaller families with early onset of Paget-like bone lesions and an autosomal dominant pattern of inheritance have been reported. One family of German origin has three affected males in two generations and was described as an osteolytic-expansive type of familial Paget's disease.[14,15] A North American family has six affected individuals in three generations (M. Whyte, personal communication). Both families show apparent autosomal dominant inheritance of early onset osteolytic lesions although the severity of their disorder is towards the lower end of the range found in the Northern Irish family. Three Italian sibs with longstanding polyostotic Paget's disease were reported in 1991,[16] two of whom died after developing an aggressive osteosarcoma. This family came from a region of Italy near Avellino. Several related patients with ancestry from this region have developed Paget's disease of bone complicated by a giant cell tumor which is amenable to treatment with dexamethasone.[17] This complication has not been found in FEO and the etiology of their disease may be different. In 1972, McKusick reviewed the inheritance of Paget's disease of bone in families with more than one affected member and concluded that the pattern of inheritance was autosomal dominant with variable clinical expression.[18]

MAPPING THE FEO GENE

It is usually essential to identify the gene responsible for an inherited disorder or its product before the mechanism of action can be established and a fully rational treatment initiated. The genetic basis of many of the bone dysplasias is unknown. Several structural proteins of bone and regulatory proteins associated with bone metabolism are not yet associated with any disorders and remain as candidates for causing inherited bone problems.

The large Northern Irish family was investigated for linkage to several of these candidate genes, including osteonectin, osteopontin, bone morphogenetic proteins and various collagen genes. These were excluded as the cause of FEO and the search for linkage was extended to cover the whole genome. Sixty-one members of generations III, IV and V were involved in this study. Diagnosis was based mainly on a clinical history of bone pain and on radiological examination, and to a lesser extent on biochemical, hearing and dental abnormalities. Four young adults in generations IV or V with generalized skeletal abnormalities but no focal bone lesions were assigned risks of 75 or 90%. More than 200 restriction fragment length polymorphisms (RFLPs) and 100 highly informative microsatellite polymorphisms were typed in this family and the FEO gene was excluded from >95% of the linkage map before linkage was established to chromosome 18q.[8] The FEO gene showed tight linkage to D18S64, D18S60, D18S55 and D18S42. There was no evidence of recombination with any of these markers, but four recombinants were identified with D18S35 and five with D18S61 which flanked the disease locus. The maximum lod score of 11.5 was obtained at a recombination value of 0 with D18S64.

When linkage was established early in 1993, chromosome 18 was very poorly represented with polymorphic markers and mapped genes. Several new markers have been identified recently and the improved map of this chromosome (Fig. 8.4) has allowed the localization of the gene to be refined. Part of the pedigree with the results of typing several polymorphic markers in the vicinity of the FEO gene is shown in Figure 8.5. The two closest flanking recombinants occur in children below the age of onset of focal bone lesions. Diagnosis of their affection status is likely to be 95% accurate. V1 has inherited the alleles of D18S383 and D18S64 which normally segregate with the disease from his affected father but is not thought to be affected, therefore placing the FEO gene distal to D18S64. The position of recombination in V1 can be refined further to between D18S64 and D18S38. V6, who is likely to have inherited his mother's FEO gene, has not inherited her at risk alleles at D18S51 and more distal markers, therefore placing the FEO gene proximal to D18S51. These two children suggest that the FEO gene maps between D18S64 and D18S51 with a certainty of >90%. Recombination in III3 and IV1, who are adults with unambiguous diagnosis, gives a definite maximum region for the gene between D18S383 and D18S483 at 18q21.2-21.3.

The distance between D18S64 and D18S51 is reported to be between 7 and 15 cM.[19-21] Identification of only 2 recombinants in 63 meioses in this family suggests that this may be an overestimation. Probable mistypings have been identified in the CEPH mapping data and each error adds approximately 2 cM to the map distance. The ratio of female:male recombination in this region in the FEO family is 3.7:1.

During the last year we have developed a YAC contig which extends from D18S483 across the probable FEO gene region to beyond marker WI-934 which is proximal to D18S64. This was achieved by hybridization of high density filters of inter-alu PCR products from the CEPH library of large insert megaYACs with inter-alu PCR probes, PCR-based screening of pooled YAC DNA with microsatellites and sequence tagged sites (STSs) within the region, and interrogation of Genethon YAC data available by Internet. The large size of the YACs is an advantage for contig formation, but they suffer from a high rate of chimerism and re-arrangement. We have eliminated YACs which are known to be highly chimeric from the contig which has at least 4-fold depth in most regions and encompasses the highly polymorphic markers D18S483, D18S68, D18S42, D18S55, GATA26C03, D18S51, D18S60, D18S499 and D18S64. We are currently constructing a restriction map of this contig which is approximately 5-6.5 Mb in

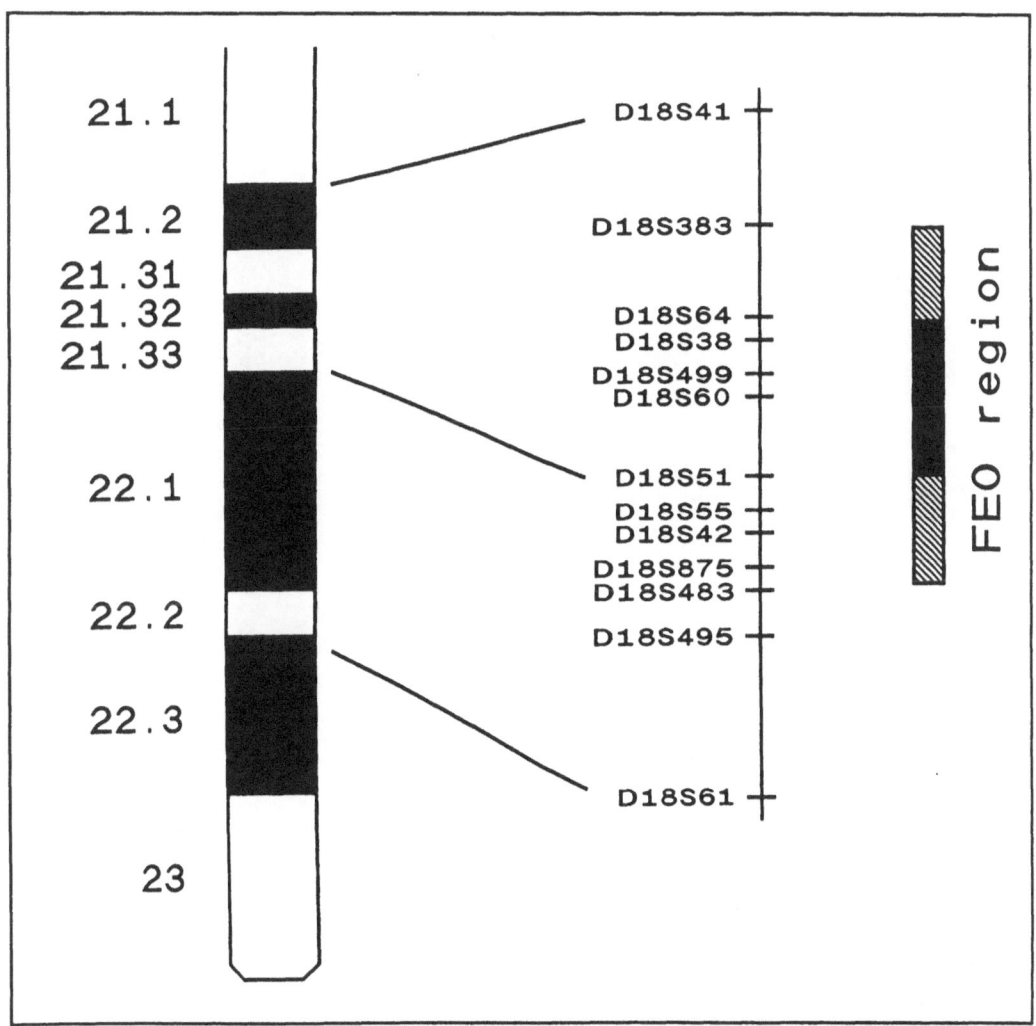

Fig. 8.4. Map of genetic markers in the FEO gene region on chromosome 18q.

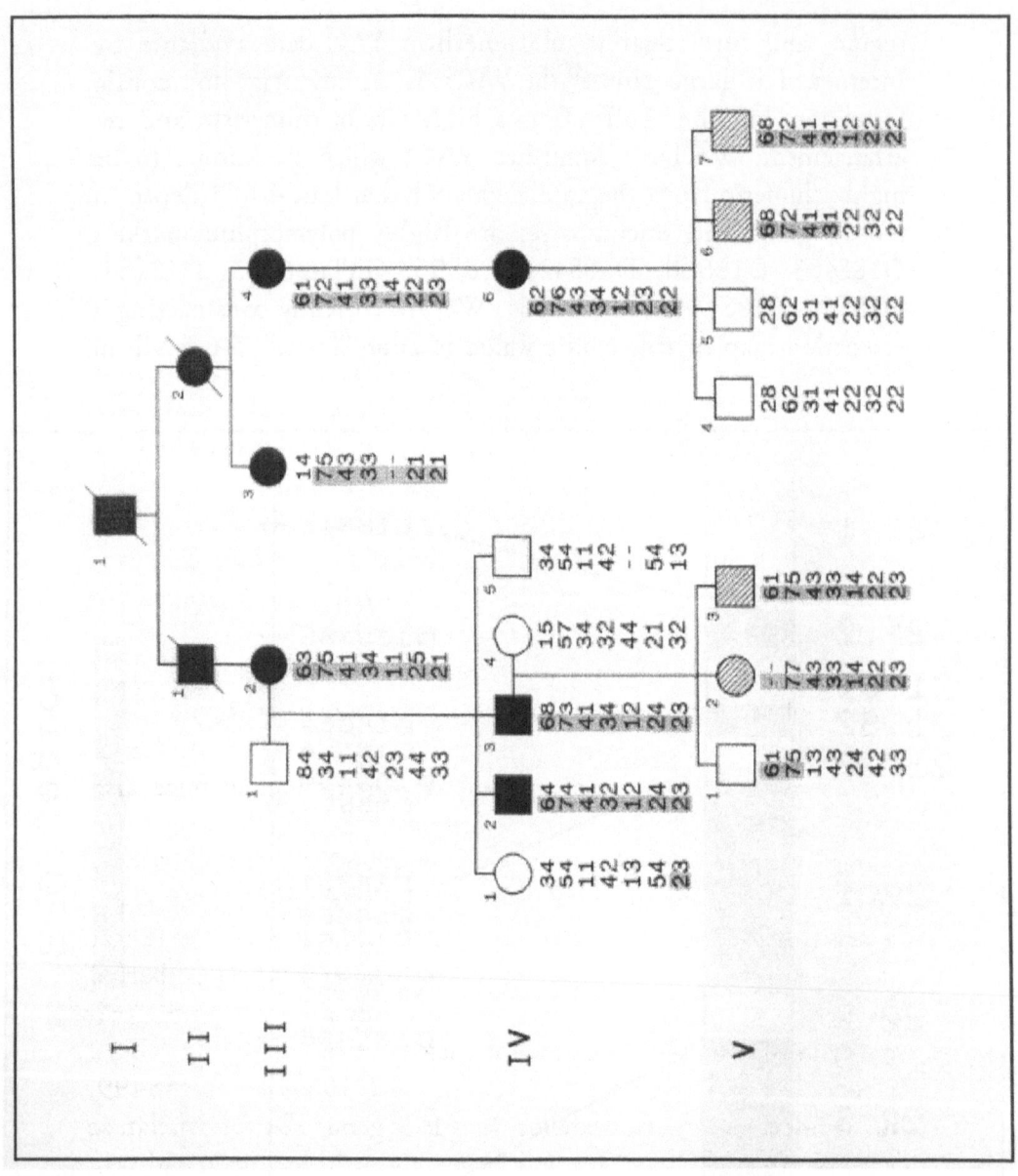

Fig. 8.5. Pedigree of part of the FEO kindred showing types for markers D18S383, D18S64, D18S499, D18S60, D18S51, D18S42 and D18S483 in order from top to bottom. Alleles transmitted from the affected parent are shown on the left of each pair and the haplotype linked to the disease gene is shaded. Analysis of recombination suggests that the FEO gene lies between D18S64 and D18S51.

length. The physical distance between D18S64 and D18S51 is estimated to be about 4.5 Mb. This contig will be used as a basis for further refinement of the FEO locus and identification of expressed sequences which may be candidates for the FEO gene.

CANDIDATE GENES

GENES AT 18Q21.3

There are few known genes located at 18q21.2-21.3 and none is a good candidate as the cause of FEO. The plasminogen activator inhibitor type-2 (PAI2) and BCL2 proto-oncogene lie on a 2 Mb YAC contig[22] which overlaps towards the telomeric end of our contig. Preliminary results suggest that the PAI2 gene is distal to D18S51, and therefore outside the probable region of the FEO gene. The BCL2 gene acts in an antioxidant metabolic pathway aimed at eliminating oxygen free radicals which can induce lesions in DNA, lipids and proteins. It also modulates intracellular Ca^{2+} fluxes, and with others (e.g., c-myc and p53) is involved in tight control of cellular proliferation and programmed cell death.[23] Rearrangements in this gene are associated with B cell lymphomas but not with skeletal problems, and transgenic mutant mice also show no skeletal abnormalities,[24] therefore the gene is unlikely to be responsible for FEO. The BCL2 gene is 600 kb centromeric to PAI2, but may also be distal to D18S51. The melanocortin 4 receptor (MC4-R) gene maps to the region just proximal to D18S51 and is within the FEO critical region. This gene is a member of the transmembrane G-protein receptor family and is expressed primarily in the brain.[25] The gene for gastrin releasing peptide (GRP) is very close to D18S95 in the vicinity of D18S64 and D18S499, but is not a good candidate for the FEO gene. The ferrochelatase (FECH) gene which is involved in the haem biosynthetic pathway and is mutated in erythropoietic protoporphyria maps proximal to D18S383 and therefore outside the FEO gene region.

In the absence of suitable candidates, it will be necessary to identify sequences from the FEO region which are expressed in bone for evaluation as the cause of FEO. This may be less difficult than expected for a region of several megabases. Chromosome 18 is composed entirely of G or mundane R bands and can be expected to have a gene density of less than 10% of the gene-rich T band regions.[26] CpG island density is also a good indicator of generalized gene density and is exceptionally low on this chromosome.[27] The tissue-specific nature of the FEO gene and the low CpG island frequency in the region of the gene suggest that a combination of bone cDNA screening and exon-trapping using YACs or cosmids is likely to be the best strategy for isolating this gene.

A MOUSE MODEL FOR FEO

Analysis of animal models can provide invaluable information to increase our understanding of human disease. In 1991 a mutation was described in the mouse[28] which is a possible model for FEO. Affected mice develop kinks in their tails at around 6-8 weeks of age, and within the next few weeks the hind feet and front paws become swollen and then severely deformed. These animals are crippled but have a normal life span. The mutation was named *cmo* because of phenotypic similarities with the human disorder chronic recurrent multifocal osteomyelitis (CRMO).[29] This is a very painful inflammatory disorder which affects multiple bone sites, often recurrently, in young children. The disease is self-limiting and patients usually recover fully. CRMO is extremely rare, and although the etiology is unknown, is not thought to have a genetic basis.

The *cmo* mutation has been mapped to mouse chromosome 18 in the interval distal to the *Adrb-2* locus and proximal to *Mbp*. Mouse chromosome 18 shows extensive linkage homology with human chromosomes 5 and 18.[30] In the mouse, there is complex disruption of the human chromosome 18q linkage group by the human chromosome 5q linkage group. This could be explained by a translocation followed by an inversion. The human homologue of *Adrb-2* maps to human chromosome 5, but the homologue of *Mbp* maps to human chromosome 18q22. *Grp*, *Fech* and *Dcc* map close to *cmo* between *Adrb-2* and *Mbp* in the mouse, and their human homologues map to chromosome 18q21 near to the FEO gene, so the location of the *cmo* gene suggests that it may be an appropriate candidate for the FEO gene. The *cmo* trait shows autosomal recessive inheritance in mice, whereas FEO is transmitted as a dominant disorder. Nevertheless, it is possible that different mutations within the same gene could show a different pattern of inheritance.

It is difficult to compare the lesions in the *cmo* mouse with those in FEO, CRMO and Paget's disease of bone and to assess whether this mouse is a suitable animal model for any of these diseases. The time course of the disease in the *cmo* mouse may suggest more similarity with FEO or Paget's disease of bone than with CRMO. The *cmo* gene has been mapped in relation to two relatively distant flanking genes. Recent advances in mouse mapping should enable a much more precise location to be identified and to establish whether its human homologue maps to within the critical FEO region. Although there can often be difficulties comparing the disease features, mouse models of human disease offer significant advantages in the areas of mapping and identification of genes. The shortage of affected families and the large physical distance between the closest recombinants flanking the FEO gene in the large Northern Irish family increase the difficulty of isolating the FEO gene. It may prove to be a much simpler task to isolate the *cmo* gene, identify its human homologue and assess the role of this gene in FEO.

VIRAL-RELATED GENES

FEO has a definite genetic basis controlled by a dominant gene located on chromosome 18, however, the late onset of the focal bone lesions and the finding of nuclear inclusion bodies typical of viruses in osteoclasts suggest that a viral trigger may be required for clinical expression. Similar nuclear inclusions are detected in pagetic osteoclasts, and viruses are now known to be implicated in the development of this disease (chapter 4).

Two retroviral sequences SSAV1[31] and ERV1[32] map to chromosome 18q21 and 18q22-qter, respectively, near to but outside the FEO gene region. Both contain a long terminal repeat-like structure with several possible regulatory elements. These sequences may have a role in human neoplasia, but their location precludes a central role in FEO.

OVERVIEW

Clinically, FEO and Paget's disease of bone have many features in common. The osteolytic lesions are very similar histologically. Nuclear inclusions suggestive of viral infection are found in osteoclasts in both conditions. Sarcomatous transformation ensues in a small percentage of lesions in either disease, although this probably occurs as a result of greatly increased cellular activity in the affected bone rather than from the disease process itself. The distribution of lesions is somewhat different, and this is hard to explain if FEO is indeed an inherited condition at one end of a wide spectrum of Paget's disease of bone. FEO shows an autosomal dominant inheritance, whereas Paget's disease of bone is usually a sporadic condition. Any genetic influence in Paget's disease is hard to study because of the late age onset, complicated by variable expression and the possibility of incomplete penetrance. On rare occasions when more than one member of a family is affected, autosomal dominant inheritance with variable expression is suggested. All families with early onset of Paget-like lesions show autosomal dominant inheritance. Several disorders (e.g., retinoblastoma) can occur either in an inherited form or as a sporadic event, resulting from mutations within the same gene. The familial cases involve inheritance of a mutant allele on one chromosome with focal disease arising after somatic loss (usually due to deletion) of the second unaffected allele. Sporadic cases require two independent somatic mutations in a single cell to knock out both normal alleles and usually develop a single lesion at a later age of onset. This two-hit mechanism could apply to Paget's disease of bone. We might then expect patients with 18q-syndrome involving deletion of the FEO gene region on one of their chromosomes to develop features of FEO. These individuals, all of whom are mentally retarded but have a normal life span, are reported to have sensorineural deafness but no skeletal problems. It is possible that homozygous deletion of additional important genes near to the FEO locus makes any somatically deleted cells nonviable, therefore preventing the occurrence of focal bone lesions.

Further research is required to assess the role of the FEO gene in Paget's disease of bone. The FEO critical region is now sufficiently small and well represented with polymorphic markers to allow association studies to be performed on DNA from pagetic patients. Sib-pair analysis and studies of Paget's families are problematic because of the difficulty collecting sufficient cases. Association studies are of limited value in large mixed populations if several independent mutations have arisen in a disease gene, but are more successful in smaller in-bred communities. If Paget's disease of bone and FEO can be confirmed within the same spectrum of disease, identification of the FEO gene and establishment of its role in the regulation of bone metabolism may help to elucidate the mode of action of these disorders. This may also lead to a more rational and effective treatment.

Identification of the FEO gene may be difficult because of the shortage of affected families. It will be possible to locate sequences from within the FEO critical region that are expressed in bone, but strategies for positive identification of the gene will depend on the size and complexity of the candidates. Cloning the *cmo* gene in the mouse model may be advantageous if this gene can be shown to map to the region homologous to FEO. The Tax mouse also shows many signs associated with Paget's disease and FEO. This model implicates viral activation of cytokine genes in the etiology of these disorders, and searching for homology between candidate genes and cytokine and other known genes may give a clue to function. It may be necessary to produce knockout mice or to use more sophisticated gene targeting methods to introduce mutations in mice to identify conclusively the FEO gene and examine its function.

Familial expansile osteolysis is a very rare inherited bone disorder with features in common with Paget's disease of bone. Studying the genetic and molecular basis of this disorder should lead to a better understanding of bone metabolism and may have important implications in Paget's disease of bone.

REFERENCES

1. Osterberg PH, Wallace RGH, Adams DA et al. Familial expansile osteolysis (a new dysplasia). J Bone Jt Surg [Br] 1988; 70B:255-260.
2. Crone MD, Wallace RGH. The radiographic features of familial expansile osteolysis. Skeletal Radiol 1990; 19:245-250.
3. Kanis JA. Epidemiology in Pathophysiology and Treatment of Paget's Disease of Bone. London: Martin Dunitz 1991:1-11.

4. Kaplan FS. Paget's disease of bone: exploring the questions. Calcif Tissue Int 1992; 51:1-3.
5. Jaffe HL. Paget's disease of bone. Arch Pathol 1933; 15:83-113.
6. Dickson GR, Shirodaria PV, Kanis JA et al. Familial Expansile Osteolysis: a morphological, histomorphometric and serological study. Bone 1991; 12:331-338.
7. Mitchell CA, Kennedy JG, Wallace RGH. Dental abnormalities associated with familial expansile osteolysis: a clinical and radiographic study. Oral Surg Oral Med Oral Pathol 1990; 70:301-307.
8. Hughes AE, Shearman AM, Weber JL et al. Genetic linkage of familial expansile osteolysis to chromosome 18q. Hum Mol Genet 1994; 3:359-361.
9. Affolter C. Caring in the community for a family with a familial bone disease. Health Visitor 1990; 63:228-229.
10. Osterberg PH. An unusual familial disorder of bone: response to calcitonin. In: Kanis JA, ed. Bone Disease and Calcitonin. Armour Pharmaceutical Company Ltd 1976:181-185.
11. Wallace RGH. A study of a familial bone dysplasia. MD thesis, The Queen's University of Belfast, Northern Ireland, 1987.
12. Barr RJ. A linkage and therapeutic study of Familial Expansile Osteolysis. MD thesis, The Queen's University of Belfast, Northern Ireland, 1993.
13. Smith R. Paget's disease of bone. J Bone Joint Surg [Br] 1995; 77-B:673-4.
14. Enderle A, Willert H-G. Osteolytic-expansive type of familial Paget's disease. Path Res Pract 1979; 166:131-139.
15. Enderle A, von Gumppenberg S. Osteitis deformans (Paget)—oder eine tarda-form einer hereditaren hyperphosphatasie. Arch Orthop Traumat Surg 1979; 94:127-134.
16. Wu RK, Trumble TE, Ruwe PA. Familial incidence of Paget's disease and secondary osteogenic sarcoma. Clin Orthop Rel Res 1991; 265:306-309.
17. Jacobs TP, Michelsen J, Polay JS et al. Giant cell tumour in Paget's disease of bone: familial and geographic clustering. Cancer 1979; 44:742-747.
18. McKusick VA. Paget's disease of the bone. In: Heritable Disorders of Connective Tissue. 4th ed. St. Louis: CV Mosby Co., 1972: 718-723.
19. Straub RE, Speer MC, Luo Y et al. A microsatellite genetic linkage map of human chromosome 18. Genomics 1993; 15:48-56.
20. CHLC maps (ftp.chlc.org).
21. van Kessel AG, Straub RE, Silverman G et al. Report of the second International workshop on human chromosome 18 mapping 1993. Cytogen Cell Genet 1994; 65:141-165.

22. Silverman GA, Jockel JI, Domer PH et al. Yeast artificial chromosome cloning of a two-megabase-size contig within chromosomal band 18q21 establishes physical linkage between BCL2 and plasminogen activator inhibitor type-2. Genomics 1991; 9:219-228.

23. Lam M, Dubyak G, Chen L et al. Evidence that BCL-2 represses apoptosis by regulating endoplasmic reticulum-associated Ca^{++} fluxes. Proc Natl Acad Sci USA 1994; 91:6569-6573.

24. Nakayama K, Nakayama KI, Negishi I et al. Targeted disruption of Bcl-2 alpha beta mice: occurrence of gray hair, polycystic kidney disease and lymphocytopenia. Proc Natl Acad Sci USA 1994; 91:3700-3704.

25. Gantz I, Miwa H, Konda Y et al. Molecular cloning, expression, and gene localization of a fourth melanocortin receptor. J Biol Chem 1993; 268:15174-15179.

26. Holmquist GP Chromosome bands, their chromatin flavors, and their functional features. Am J Hum Genet 1992; 51:17-37.

27. Craig JM, Bickmore WA The distribution of CpG islands in mammalian chromosomes. Nature Genet 1994; 7:376-382.

28. Byrd L, Grossman M, Potter M et al. Chronic multifocal osteomyelitis, a new recessive mutation on chromosome 18 of the mouse. Genomics 1991; 11:794-798.

29. Giedion A, Holthusen W, Masel LF et al. Subacute and chronic "symmetrical" osteomyelitis. Ann Radiol (Paris) 1972; 15:329-342.

30. Justice MJ, Gilbert DJ, Kinzler KW. A molecular genetic linkage map of mouse chromosome 18 reveals extensive linkage conservation with human chromosomes 5 and 18. Genomics 1992; 13:1281-1288.

31. Brack-Werner R, Barton DE, Werner T et al. Human SSAV-related endogenous retroviral element: LTR-like sequence and chromosomal localization to 18q21. Genomics 1989; 4:68-75.

32. O'Brien SJ, Bonner TI, Cohen M et al. Mapping of an endogenous retroviral sequence to human chromosome 18. Nature 1983; 303:74-77.

INDEX